Self-Organisation Shapes Travel Behaviours and Social Exclusion in Deprived Urban Neighbourhoods of China

Joseph Cho-yam Lau

Self-Organisation Shapes Travel Behaviours and Social Exclusion in Deprived Urban Neighbourhoods of China

Springer

Joseph Cho-yam Lau
University of Hong Kong
Hong Kong, China

ISBN 978-981-15-2251-2 ISBN 978-981-15-2252-9 (eBook)
https://doi.org/10.1007/978-981-15-2252-9

This Springer imprint is published by the registered company Springer Nature Singapore Pte Ltd.
The registered company address is: 152 Beach Road, #21-01/04 Gateway East, Singapore 189721, Singapore

Contents

List of Figures

List of Tables

Chapter 1
Introduction

1.1 Introduction

Since 1978, China has transformed from a central planning economic system to an open market economy system, and a great number of industrial workers have been laid off from state-owned enterprises (SOEs) and moved to newly established labour markets in this process. As a result, many unskilled and low-income workers have been excluded from accessing decent housing and formal employment. In a deprived inner city in Guangzhou, Xiguan, after 40 years of economic and housing reforms, many low-income residents still have to live in dilapidated, deprived and overcrowded housing conditions. When the Chinese Communist Party obtained control of China in 1949, many rich residents left China, and the Guangzhou municipal government allocated vacant houses in Xiguan to workers through work units. A high percentage of the existing residents moved in during that period because the government sold houses to them at discounted prices during the housing reforms in the 1990s (Yuan et al. 2011).

Most of the residents in this inner city are redundant workers of SOEs. According to the usual practices of urban redevelopment, the municipal government has adopted a privately funded and property-led redevelopment approach to redevelop the inner city. However, most private property developers are reluctant to participate in the redevelopment of the inner city due the high employee compensation costs, low profits and the fact that the developers required to preserve the historic houses. Lacking redevelopment, many large houses have turned into dilapidated and dangerous buildings. Residents must face congested, narrow streets and conflicts between traffic and pedestrians. Due to their low educational attainment and poverty, the residents cannot move out of the city; they can only find informal jobs along the tourist walking streets that are developed by the economic revitalisation policy. As a result, local residents have become accustomed to riding bicycles to both work and non-work activities since the *danwei* era, and they commute shorter distances. The

J. C.-y. Lau, *Self-Organisation Shapes Travel Behaviours and Social Exclusion in Deprived Urban Neighbourhoods of China*,
https://doi.org/10.1007/978-981-15-2252-9_1

residents are excluded from participating in the mainstream activities of society (Zhou et al. 2015; Yuan et al. 2011). They face accessibility problems in reaching employment, social and recreation activities that are further away.

Furthermore, the market reforms have absorbed millions of rural migrants into the cities. In 2016, approximately 287 million rural migrant workers found employment in urban China, making up more than one-third of the entire working population (China Labour Bulletin 2017). Many rural migrants have been excluded from accessing social services and formal job markets in cities because they hold rural household registration (*hukou*) status. The migrants cannot exchange their labour power for formal jobs to maintain their quality of life. Given the constraints imposed by the *hukou* system, most rural migrants' freedom and choices in seeking jobs and housing are spatially restricted to the inner suburbs of the host cities (Shen 2017; Zhu 2016). The migrants gather in urban villages or migrant enclaves and depend on effective the social networks among fellow villagers to find informal jobs and affordable housing.

In addition, in face of their farmland being acquired by the government, local villagers in urban villages make use of the dualistic land-use system to build affordable housing and township-and-village enterprises on rural land in order to tap into the cheap migrant labour force. The actions of villagers and rural migrants produce a jobs-housing co-location process in order to cut travel times (Zhu et al. 2017; Zhao and Howden-Chapman 2010). The migrants' daily activities are almost completely spatially restricted to the enclaves. They take short-distance trips and use slow transport modes to get to work and non-work activities. They are excluded from participating in the mainstream activities of society.

In addition, a study on the suburbanisation of Shanghai found that the Line 9 rail transit services, which link the suburban district, Songjiang, and the central city, lacks sufficient ridership to support cost-effective rail transit services (Pan et al. 2011). Government transport planners followed the conventional four-step urban transport planning model and the design guidelines of transit-oriented development (TOD) to build the rail transit system in the suburbs. Thus, the declining ridership of the rail services in Shanghai is not due to the forecasting problems of conventional urban transport planning models (Wang et al. 2012; Pei and Fan 2010; Meyer and Miller 2001). There are likely other factors that are outside the framework of conventional transport planning models that cause declining rail ridership problems.

The above three cases in urban China showcase typical commuting and social exclusion problems in conventional transport studies. The travel behaviours that have contributed to commuting problems and social exclusion cannot be fully measured by conventional urban transport planning. These methods of planning always focus on the influence only of the built environment and socio-economic factors on the travel behaviours of the people. For example, transport planners in Shanghai were unable to accurately forecast rail travel demand, resulting in the oversupply of rail services. In fact, social exclusion and commuting problems can be better resolved by understanding the interaction between individuals' travel behaviours and government policy (Giddens 1984). For example, in the cases of Xiguan and urban villages, government policy-makers have failed to observe the decisions and actions of rural migrants and low-income workers when interacting with government policies, resulting in social exclusion and accessibility problems.

Human social systems are complex social systems. Travel behaviour is the product of spontaneous and non-linear self-organisation processes (or interactions between structure and individuals' actions). Structure shapes action, but individual actions produce and transform structure to form new travel behaviours (Giddens 1984). The new behaviour is called emergence, in which transport planners cannot depend on using structural factors (e.g., the built environment) alone to predict the future trends of changing behaviours (Morrison 2005; Fuchs 2003; Stacey 1996). As a result, conventional transport planning and forecasting always result in uncertainties and discrepancies between actual and forecasted travel behaviour (Andersson et al. 2017; Xia et al. 2016; Blumenberg 2004).

Hence, to fill this research gap, this book introduces a new transport planning approach called the self-organisation process, which will help us to investigate travel behaviour. A self-organisation process refers to the interaction between structures and the individual actions of actors in social systems. The process is able to help transport planners forecast the trends of new travel behaviours, as forecasting demands and providing solutions to transport problems seem to be difficult tasks for conventional urban transport planning models. As a result, city decision-makers can obtain more information that is accurate for supplying transport infrastructure to match future travel demands.

The self-organisation approach can help transport policy-makers identify the trends of new travel behaviours (e.g., low-income workers making short-distance trips in urban China) to understand the causal factors that result in commuting problems and social exclusion. The approach can also help transport policy-makers use social justice theories and public consultation approaches to deliver equitable policies for minimising social exclusion and improving accessibility.

1.2 Self-Organisation Approach Redefines Travel Behaviour

Self-organisation is a process that takes place in response to external adversity in a social environment. It is a natural and spontaneous mechanism of interactions among members of a social system in an effort to achieve a state of equilibrium. For example, when swarms of honeybees want to avoid danger, they move together, and sometimes, the swarm behaves as if it were a single giant animal. In addition, self-organisation is seen in the free market system. As early as 1776, Adam Smith's 'invisible hand' referred to a self-organising mechanism guiding and shaping economic systems (Smith 1937). Krugman (1996) illustrates the economic consequences of self-organisation: 'Self-organisation is something we observe and try to understand, not necessarily something we want' (Krugman 1996, pp. 5–6).

Recently, the concept of the self-organisation process has been used in urban planning. Batty (2012) redefines urban regeneration within the context of a wider system of urban reproduction. The literature on urban regeneration is usually structured through planned interventions that are anything but spontaneous and often conflict with processes that are intrinsic to the survival of the system itself. Batty redefines urban regeneration as the interaction between the planned interventions and spontaneous actions of the residents. Urban areas are living and self-organising systems that are composed of entities—people and buildings—that have limited lifespans and have to be renewed continuously.

For example, a study in Nanluoguxiang, an inner city in Beijing, demonstrated that urban regeneration is itself a self-organisation process (Zhang and De Roo 2016). In the 1950s, as a remedy for a housing supply shortage, the Beijing government developed the 'Jingzu policy', which transferred the ownership of the 'siheyuan', or courtyard houses, from private hands to the government, and the houses were subdivided into small rooms and leased to many low-income tenants. The pressure of the increasing population reached a critical point, prompting residents in the inner city to carry out spontaneous self-organised actions, e.g., they carried out self-building activities to create extra individual rooms in courtyards, that were in line with the Jingzu policy. The policy therefore started the self-organisation process. Finally, the process produced new structural properties that transformed the 'siheyuan' into overcrowded homes and turned the inner city into a tourist destination.

The process of self-organisation in social systems produces new structural properties (e.g., new behaviours), a phenomenon that is called emergence (Fuchs 2003). Emergence refers to a new order emerging from an old system in phases of instability and crisis in order to restore the system to a balanced state. In this study, emergence is the product of an interaction between individual actions and government rules, and the interdependent relationship between the two variables comes from the fact that the structural factors are the medium and outcome of individual travel behaviours (Fuchs 2003; Giddens 1984). Hence, emergence or new transport behaviours cannot be forecast by conventional urban transport planning models. The self-organisation process provides new travel behaviour trends so that transport planners can understand and forecast future travel demands to provide appropriate transport systems. For example, when travellers drive cars in a busy city, structural factors (e.g., transport infrastructure) influence travellers to travel on road networks, and at the same time, travellers transform the transport infrastructure through the reproduction of their routine driving practices. The interactions between driving actions and structures produce new travel behaviours. Transport planners can predict the trends of drivers' new travel behaviours by means of the self-organisation approach.

Self-organisation is an evolutionary feedback loop (feedback loops are a process whereby a change to a social system results in an alarm that will trigger a certain result within that same system) that occurs in social systems. The process yields new structural properties (e.g., travel behaviour) to change the environment (e.g., transport system) over time. The structuration theory (Giddens 1984) is used as the mechanism of the self-organisation process, and the concepts of self-organisation, in particular, the idea that social systems can be continual self-producing, place the production and reproduction of social systems at the centre of their theories (Morrison 2005; Fuchs 2003).

1.3 Structuration Theory Is the Mechanism of the Self-Organisation Process

The structuration theory is used as the mechanism of the self-organisation process to investigate commuting practices and social exclusion. Giddens (1984) suggests that structuration theory describes the interaction between structure (e.g., rules and resources) and agency (e.g., practices). The concept of structuration theory is that structures shape an individual's practices and that individuals are not passive objects. Individuals show active agency in transforming structures and the environment through the reproduction of routine practices to achieve their intended goals. Human actors' individual memories store social norms, values and rules that shape their routine practices. Thus, individuals know what they are doing and how to do it in routine social activities. Therefore, Giddens states that structures are 'both the medium and the outcome of the practices which constitute social systems' (Giddens 1981, p. 27).

For example, a toll bridge is built to connect urban areas and islands to improve commuting between housing and employment centres. The policy-makers originally want to use the bridge tolls to cover the building costs of the bridge. The self-organisation process starts when the drivers, who live on the islands, draw upon their knowledge of the traffic rules to drive on the roads, pay the tolls and get to their jobs via the bridge to where the employment opportunities in the city centre are. In this situation, traffic rules are the medium for driving practices. Freedom and the availability of choices are a prerequisite for a person to execute his or her practices. Drivers make individual decisions to transform traffic rules through the reproduction of their routine practices. Some drivers choose to change their transport modes and take rail transit services instead of driving cars to the urban areas to avoid paying the bridge tolls. These new commuting practices are called new structural properties in the self-organisation process (Fig. 1.1).

Emergence includes bottom-up emergence and top-down emergence. The bottom-up emergence refers to changes that commence because of the interaction of the human actors of the social system. The top-down emergence refers to what is called structure, and it refers to changes that come from government levels (Fuchs and Hofkirchner 2005). Furthermore, the self-organisation approach helps policy-makers transform unjust policies that cause social exclusion and commuting problems. The approach is based on the concept that policy is both the medium and the outcome of the action, and thus policy and individual actions are two interdependent variables. Policy decision-makers, who want to improve people's quality of life, can

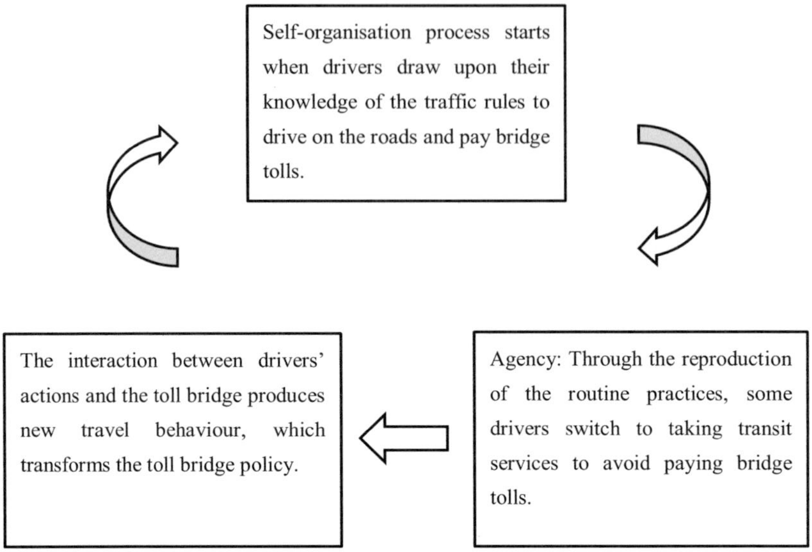

Fig. 1.1 Self-organisation process and travel behaviour

change either policy or individual actions to minimise social exclusion and resolve commuting problems.

1.4 Self-Organisation Approach and Transport Planning

Transport planning is crucial for the strategic development of transport systems and decisions about investment in transport infrastructures. However, most of the transport planning models are based on information that comes from structural factors alone to forecast travel behaviour, and the models ignore the impacts of individual actions and attitudes on travel behaviour. These transport-planning models are bound to make mistakes. This is because structure (e.g., transport rules) is both the medium and the outcome of individual actions (Giddens 1981). That is, travel behaviour is shaped by transport rules, but transport rules are the product of travel behaviours. Travellers modify transport rules through the reproduction of travel practices to achieve their goals, and at the same time, the practices produce new travel behaviours. Therefore, trends in new travel behaviours and future travel demands are constituted by travellers' practices and social structures.

For example, one transport study tried to identify a stagnated and even declining trend of car ownership (the 'peak travel' phenomenon) and travel demand in Hong Kong (Zhou and Wang 2018). The study depended on the socio-economic information of travellers (structural factors) to predict the travel behaviour of the drivers. The results of the study indicated that the percentage of households with one or more cars decreased from over 13% in 2002 to approximately 10% in 2011, suggesting that households with similar socioeconomic backgrounds were less likely to own a car in 2011 than in 2002. Discrepancies exist between the above transport forecast and the actual travel behaviour of drivers in Hong Kong. In 2018, there were approximately 2,264,000 full driving license holders, and this figure represented a growth of 25.8% compared to the numbers in 2007. In addition, the number of licensed private cars was initially 421,062 vehicles in 2008, which increased to 494,646 vehicles in 2012 and reached 660,443 vehicles in 2017 (Hong Kong Transport Department 2018). Furthermore, the average mechanised trip rate in Hong Kong was initially 1.82 trips per person in the 2002 Travel Characteristics Survey (TCS) and increased to 1.83 trips per person in the 2011 TCS (Hong Kong Transport Department 2014). The forecast of the above study could have been improved if it had observed the influence of the individual attitudes and actions of the drivers.

The arguments of this book are that transport rules, transport systems and the built environment shape the daily travel practices of travellers. At the same time, travellers transform these structural factors through the production and reproduction of their routine commuting practices (Giddens 1984). The self-organisation process produces new behaviours or emergence, which are complex and spontaneous results and thus are difficult to forecast using conventional transport planning models (Gershenso 2018; Portugali 2000, 2006; Morrison 2005). The self-organisation process is regarded as a new transport planning approach that can provide information on trends of new travel behaviour (e.g., increasing trip rates and car ownership in Hong Kong), which transport planners can use to forecast future travel demands and provide solutions for commuting problems. Hence, for better forecasting travel behaviours that are produced in an open and complex human social system, self-organisation is an ideal tool to investigate and analyse new travel behaviours.

1.4.1 Self-Organisation Process Influences Accessibility

During the process of urban development in social systems, the self-organisation process, including the individual decisions and actions of actors interacting with structures in order to restore a balanced environment, takes place. Sometimes, the self-organisation process yields new travel behaviours that cause accessibility constraints, which influence individuals' commuting practices in an attempt to achieve well-being. Accessibility is a tool to evaluate the new travel behaviours that influence low-income workers' quality of life. For example, Chap. 4 of this book describes how some low-income residents in the historic inner city of Guangzhou changed their practices to take public buses instead of riding bicycles to their places of employment. The new travel behaviours also caused accessibility problems in the neighbourhood because the roads in the historic inner city were originally built for horse-carriages and rickshaws. The working residents faced heavy traffic congestion in peak hours.

Accessibility is defined as the ability of an individual to penetrate constraints to access the mainstream activities of a society. The definition involves two important concepts: the term 'ability' refers to an individual's freedom to move about and the term 'the mainstream activities of a society' is linked to capability and social exclusion (Van Wee and Geurs 2011). In regard to the relationship between human rights and accessibility, the capability approach (Sen 1984) states that a government characterised by justice and equity should provide all people with the freedom, choice and opportunities to achieve well-being, and this right includes the freedom to penetrate socio-economic and land-use constraints to commute to necessary activities (Cao and Hickman 2019; Ryan et al. 2019). People should have the freedom to choose their transport modes, travel distances and trip frequency involved in traveling to job and non-work activities. For example, a study examined the capabilities of the people who lived near three stations on the Beijing subway line 1, namely Guomao, Sihui and Tuqiao (Cao and Hickman 2019). The results of the study indicated that functionings (e.g., an individual's realised behaviour) and capabilities differ according to an individual's socio-economic characteristics and geographical location, meaning that both the opportunities to participate in life and also the actual activities themselves differ.

Furthermore, a person's accessibility depends on his or her ability to penetrate the barriers imposed on low-income workers in accessing the mainstream activities of a society. These barriers mainly include the built environment and transport systems. Therefore, in Chap. 5, there is an examination of how the residents in the inner city of Xiguan were obstructed in their attempts to commute to their places of employment as a result of the inadequate provision of public transit, conflicts between pedestrians and traffic and outdated road systems; the residents are regarded as having experienced the inaccessibility of participating in the mainstream activities of their society.

Accessibility can be used to judge the impacts of the self-organisation process on poor workers. For example, the *hukou* system in urban China interacts with the decisions and actions of rural migrants to produce new structural properties, including traveling short distances as part of their work obligations. The new behaviours directly exclude the rural migrants from participating in recreational and social activities that are located outside their living neighbourhoods.

In fact, transport planners can also evaluate the impacts of the self-organisation process on the accessibility of poor workers by means of the concept of travel needs. Travel needs refer to the needs that individuals have for transport tools and transit services in order to reach activities that satisfy their basic needs (e.g., employment, education and social participation) (Pyrialakou et al. 2016). For example, Chap. 6 traces how the policy of farmland acquisition in the suburban areas of Shanghai resulted in urban sprawl. Many displaced local workers in new towns felt unable to take public transit to job opportunities in the city centre. The land acquisition policy interacted with the actions of the displaced workers and produced new structural properties, which included the inadequate provision of public transit services and displaced workers' accessibility problems.

Furthermore, new travel behaviours that are produced by the self-organisation process can be divided into two types of trips: obligatory and discretionary trips. Obligatory trips are those trips that individuals have to undertake to fulfil their survival needs, such as trips to places of employment. Discretionary trips are those trips that individuals take to develop their personal social development and recreational needs. Chap. 4 of this book describes how, because of the interaction between the *hukou* system and individual actions, most migrant workers choose to use most of their time intervals for obligatory trips (e.g., employment and childcare), and they ignore discretionary trips (social and recreation participation). As a result, the self-organisation process produces new travel behaviours that result in commuting problems and the exclusion of rural migrants from accessing non-work activities.

1.5 Transport-Related Social Exclusion and the Self-Organisation Process

The concept of the self-organisation process is used as a tool to investigate social exclusion and commuting practices. The structural factors interact with individual actions to produce new structural properties, resulting in commuting problems. The process also contributes to the exclusion of low-income workers from participating in the basic economic and social activities of the society in which they live (Schwanen et al. 2015; Lucas 2012).

Social exclusion is defined as a process of interaction between structure and agency that produces a state in which individuals are obstructed from joining or unable to participate fully in the economic, social, political and cultural life of their society (Giddens 1984). Participation may be obstructed when people lack access to material resources, and participation is limited when people cannot exercise their voice or interact with each other and when their rights and dignity are not accorded equal respect and protection. Thus, social exclusion is a process of interaction between structure and agency that produces a state in which individuals are obstructed from joining or unable to participate fully in the economic, social, political and cultural life of their society. In addition, social exclusion is a multi-dimensional phenomenon that is not limited to material deprivation; poverty is an important dimension of exclusion, albeit only one dimension. Other dimensions in the process of urban development, such as capability deprivation and accessibility inequity, should also influence social inclusion. Measuring social exclusion is challenging due to its multi-dimensional nature; the measurement should involve access to resources, commuting inequity in meeting basic needs and participation in the mainstream activities of a society in accordance with justice, freedom, and human rights (United Nations 2016; Sen 1984, 2000; Rawls 1971).

Transport-related social exclusion not only focuses on the provisions of transport services or the abilities of people to physically move around but also involves unjust governmental policies and individual actions that constrain individuals' commute required to achieve well-being because of a lack of freedom, choice and opportunities (Schwanen et al. 2015; Lucas 2012; Preston and Rajé 2007).

Various individual actions, norms, and government rules exclude low-income workers from achieving well-being. The variables that constitute social exclusion include two types. The first type are structural factors, such as urban redevelopment policies, the *hukou* system, the gendered division of household labour and dualistic land-use policies. The second type are the individual actions of low-income workers, such as rural migrants gathering in an urban village to find jobs, and of local residents in an inner city who choose to live in an overcrowded environment.

For example, studies in Chap. 5 of this book indicate that the economic and housing reforms caused many of the low-income residents in Xiguan to become redundant SOE workers, and they fell into the trap of poverty. They could not afford the commodity housing built by the private sector, and they sub-divided their houses into several rooms so that each could house more people. Furthermore, to reduce travel costs, low-income residents made the individual decision to live near the informal jobs that were created by the revitalisation policy. In addition, these residents chose to live near the social networks that they had developed over many years in Xiguan. The results of the case studies in Chap. 5 indicate that unjust policies interacted with individual actions to produce social exclusion. In addition, the residents experienced social exclusion that was developed by a multidimensional phenomenon, such as housing deprivation, and the low-income residents could not exchange their labour power for adequate income because of the unjust policies.

1.5.1 Transport-Related Social Exclusion Model and Self-Organisation

Based on the concepts of self-organisation and accessibility, this section introduces a new transport model in urban transport planning and social exclusion. The investigations between Chaps. 4 and 7 on the deprived urban neighbourhoods in Guangzhou and Shanghai will be based on this model. The framework of the model is that the interactions between different kinds of structural factors interact with individual actions in the deprived urban neighbourhoods to produce new travel behaviours, which include poor workers' new travel distances, trip frequency and mode choices. The conventional urban transport-planning model fails to forecast these new travel behaviours because the behaviours are the product of a complex self-organisation process. The model helps transport decision-makers identify the unjust policies and individual actions that cause accessibility constraints and social exclusion, examine social justice theories to modify policies, and conduct a public consultation to understand related actions. The transport model is divided into five stages, which are as follows (Fig. 1.2):

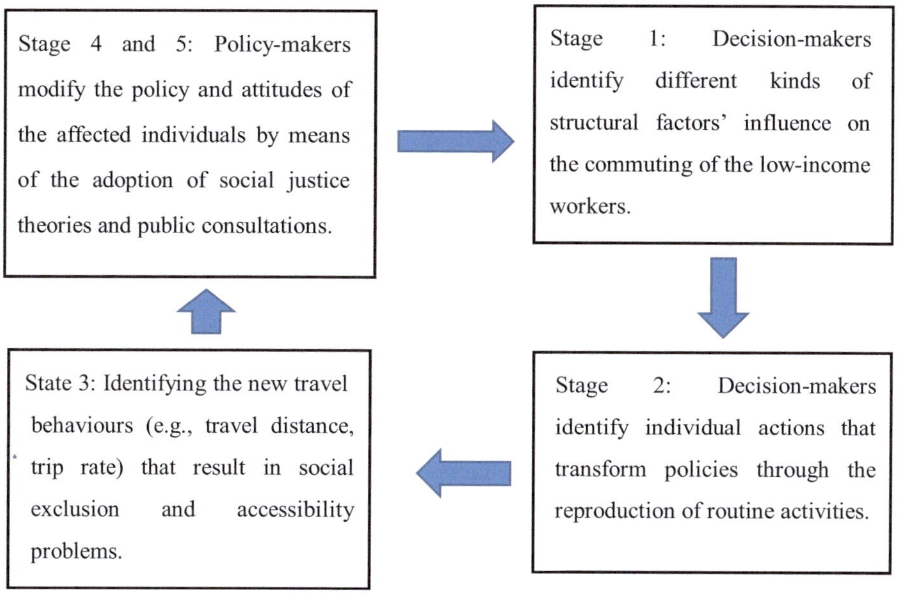

Fig. 1.2 Transport-related social exclusion model

- The first stage is to identify the structural factors that influence the commuting practices of low-income workers in their attempt to achieve well-being;
- The second stage is to identify the impacts of individual actions and attitudes of low-income workers on their inaccessibility and social exclusion;
- The third stage aims to identify new travel behaviours that cause the multi-dimensional exclusion of low-income workers;
- The fourth stage aims to identify and modify unjust policies in the self-organisation process to minimise social exclusion and improve the commuting practices of low-income workers.
- The fifth stage aims to identify the attitudes and actions of the affected individuals through the consultation process. The policy-makers can then revise the unjust policies to minimise social exclusion and improve accessibility.

The first to third stages were discussed in the previous sections. The fourth stage is based on the concept of structuration theory, in which government policy is both the medium and the product of action. In other words, structure and individual actions are inter-dependent, and policy-makers can modify new behaviours to minimise social exclusion by changing either policies or individual actions. In the fourth stage of the model, policymakers adopt social justice theories and develop possible policy responses (Fig. 1.2). This book tries to use the capability approach (Sen 1984) and the difference principle (Rawls 1971) to guide government policy-makers in making appropriate policy responses.

Stage five is based on the idea that public consultation can inform policy-makers about the needs, abilities, skills, resources, capacities and past experiences of the affected individuals during the design of a policy solution (Elvy 2014; Lucas 2012) (Fig. 1.2). This book proposes a collaborative approach as a strategy to understand the actions and attitudes of affected low-income workers. Public participation is a strategy for policy-makers to understand individual actions and attitudes so that they can modify policy or structure to enable individual actions. Healey (1997) proposes the collaborative approach as a strategy for public consultation in policy formulation. The collaborative approach should treat low-income workers and rural migrants as stakeholders and should impose a duty on all levels of the senior cadres in China to attend to the concerns of all stakeholders.

1.6 Self-Organisation Process and Social Justice

This section introduces social justice theories to help decision-makers who want to build a just society in urban China adopt social justice theories to improve accessibility and minimise social exclusion in the self-organisation process. Three social justice theories will be introduced: the utilitarian, Rawlsian and capability approaches (Pereira et al. 2017).

The utilitarian approach aims to maximise the total benefit to society. In urban transport planning, the cost and benefit analysis (CBA) in transport projects draws on the concept of the utilitarian approach by comparing the expected costs and benefits of various alternative transport policy options (Boisjoly and El-Geneidy 2017; Wee and Roeser 2013). However, the author proposes that the application of utilitarianism is not compatible with justice because it ignores the fair distribution of resources and freedom to achieve well-being. The utilitarian approach is deemed especially problematic in the *hukou* system in China. The Chinese government allows rural migrants to find jobs in coastal cities, where the migrants can provide cheap labour for industrial production. However, the government denies rural migrants access to social welfare and formal employment in the host cities.

To improve the social inclusion and commuting of migrant workers, this book suggests the Rawls' (1971) difference principle, which states that:

> Social and economic inequalities are to be arranged so that they are both: (a) to the greatest benefit of the least advantaged, consistent with the just savings principle. (p. 266)

According to the difference principle, some inequality of wealth and income is probably necessary in order to maintain high levels of productivity in China. However, social and economic inequalities are considered equitable if and only if the Chinese government makes it a top priority to help low-income workers achieve well-being through economic reforms meant to address accessibility and income and housing inequalities. Therefore, if the *hukou* system changed in accordance with the difference principle, it would allow migrants to access social services and formal jobs.

This book also proposes another social justice theory, the capability approach (Sen 1980, 1981, 1984, 1992). The capability approach entails two core claims for social justice: first, a government should provide freedom for all people to achieve well-being through such means as employment and housing. The second claim is that in order to obtain the freedom to achieve well-being, government decision-makers who want to build a just society should provide a fair distribution of capabilities for all people. In other words, each person should have real opportunities, freedom, choice to achieve well-being, including being employed, being educated, being housed in public housing, and having the mobility to access employment.

Preston and Rajé (2007) suggest that social exclusion is not caused by a lack of employment, education and health opportunities but a lack of equitable government policies to ensure that poor workers have the freedom, choice and availability of opportunities to achieve well-being.

1.7 Other Contributions of Self-Organisation to Urban Transport Planning

Other contributions of the self-organisation process to transport planning and social exclusion are as follows.

First, based on the concept of neighbourhood effects, the travel behaviours of low-income workers in deprived urban neighbourhoods should be different from those travel behaviours in other neighbourhoods. This is because the self-organisation processes between these two types of neighbourhoods are significantly different. The neighbourhood effect is a social science concept that suggests that neighbourhoods have either a direct or an indirect effect on behaviours. For example, land-use types are significant in explaining commuting time. The socio-demographic characteristics of both individuals and neighbourhoods play an important role in shaping individuals' commuting behaviours. In fact, this is the central idea of the conventional urban transport-planning model used to predict aggregated travel behaviours across different zones or neighbourhoods in a city.

Studies described in the following chapters of this book found that in the deprived neighbourhoods of Guangzhou and Shanghai in China, there were links between the neighbourhoods and social structures (such as government policies, norms, social networks) and the individual actions. Self-organisation processes that were taking place in the deprived neighbourhoods produced new individual travel behaviour trends that obstructed the low-income working residents in commuting to and participating in the mainstream activities of their society. This book argues that the impact of neighbourhood effects (e.g., the interaction between unjust government policies and the individual actions of the residents) on individual commuting behaviours may be obscured in the conventional urban transport planning models that measure the aggregated travel patterns of people across all neighbourhoods in a city. Hence, this book suggests the integration of the self-organisation model into the conventional urban transport-planning model to improve traffic forecasts and minimise social exclusion.

Within the framework of the conventional urban transport-planning model, travel behaviour forecasts have followed the conventional sequential four-step transport-planning model. The four-step model has been widely used across large cities in China (Wang et al. 2012; Pei and Fan 2010). The four steps are as follows (Bliemer et al. 2017; Andersson et al. 2017; Meyer and Miller 2001; McNally 2000):

1. The trip generation step is done to forecast the number of home-based and non-home-based trips produced from origins and attracted to destinations in each study zone. The built environment, transport system and socio-economic characteristics of the population in a zone determine the total trip generation.
2. The trip distribution step is done to forecast the origin-destination trip distributions of travel in the future made by population, employment and other demographic changes to reflect changes in people's choice of destination. Travel distance determines trip distributions. The further apart the origin and destination locations are, the fewer trips the travellers will make between them.
3. The mode choice step is done to forecast the proportion of a particular transport mode in the origin-destination trip matrix. Thus, the travel distances, trip frequency distribution per day (or trip rate), built environment, transport system and socio-economic characteristics of the population in a zone determine future mode choices; and
4. The traffic assignment step is done to assign the origin-destination trip matrix to a road network. This step spreads traffic flows and transport modes in accordance with the provision of the road network and yields travel times. The forecast is based on the idea that drivers choose the shortest travel distance or travel time to their destination.

The four-step models, which depend on the built environment and socio-economic characteristics of the population alone to predict travel behaviours, always face forecasting errors because the input variables (such as travel distance, trip frequency, travel time and mode choice) are always different in reality from what is assumed in the forecasts (Andersson et al. 2017; de Jong et al. 2007; Flyvbjerg et al. 2005).

To provide forecasts that are more accurate and minimise social exclusion in the deprived urban neighbourhoods in cities, this book will introduce steps to incorporate the results of the self-organisation transport model into the conventional four-step transport-planning model in Chap. 7.

Second, this book suggests that individual travellers generate many spontaneous travel patterns through which they transform social policies and norms through the reproduction of routine commuting practices (Giddens 1981, 1984). Hence, in the process of urban transport planning, transport planners should consider the influence of bottom-up emergence. For example, in a study on excess commuting in the US, Giuliano and Small (1993) suggest that excess commuting is presumably not agreeable to reduction through strategies for jobs-housing balance because it arises from considerations other than the minimisation of commuting costs. These considerations should include individual decisions and actions.

In addition, a transport study in Brazil indicated that gender differentials are observed for single and formerly married working women (Neto et al. 2015), which suggests that other cultural or environmental factors are not fully captured by transport studies on gender gap commuting. The results are reflective of attitudes and individual decisions regarding gender in Brazil, whereby men tend to be socially dominant and women are expected to be self-sacrificing and submissive. In fact, many transport studies demonstrate that individual decisions and actions are important factors in determining travel behaviour.

1.8 Objectives of the Book

This book has several objectives. The first objective is the application of the self-organisation process to investigate the complex travel behaviours, social exclusion and accessibility problems of low-income workers in the cities of urban China. A vast number of transport studies have reported that travel behaviour is the product of the built environment and transport integration (Pan et al. 2011, 2017). This book argues that travel behaviour is a complex travel behaviour, and thus, it is necessary to use a complex process to investigate it. This book chooses the self-organisation process. The results of the studies in Chap. 6 indicate that the interactions between structures and individual actions have produced new travel behaviours, which has led to the declining ridership of rail transit services in Shanghai.

The second objective of this book is to link the self-organisation process and the structuration theory to investigate travel behaviour. The continual interaction between structure and agency, which constitute human social systems, is a characteristic that the self-organisation approach and structuration theory share. The recursive process can be shaped in terms of the duality of structure, which means that the structural properties of social systems are both the medium and the outcome of the practices that constitute those systems (Giddens 1981). Structure and agency are a duality; neither one can exist without the other. This fundamental concept can be used to improve travel behaviour by changing policies or individual actions.

The third objective of this book is to integrate the concepts of social justice theories into urban transport planning. This book suggests that interactions between individual actions and structures produce new social system structures. The results of the studies in the following chapters indicate that unjust government policies (e.g., the *hukou* system) always result in accessibility and social exclusion problems. Individual actions and policies are interdependent variables that sometimes produce social exclusion. Thus, changing the policies will also change individual actions, improve the commuting process and minimise social exclusion. The concepts from social justice theories will help decision-makers in city governments identify and modify unjust policies deliver equitable policies. Structure is capable of both constraining and enabling action (Giddens 1984). Public participation is a strategy through which policy-makers can understand individual actions and attitudes so that they can modify policies to enable individual actions. Healey (1997) proposes the collaborative approach as a strategy to help policy-makers understand the actions of the affected individuals and deliver equitable policies for minimising social exclusion and commuting problems.

The fourth objective of this book is to apply the self-organisation approach as a tool to investigate transport-related social exclusion. According to the capability approach (Sen 1984), a government characterised by justice and equity should provide all people with the capability to participate in the mainstream activities of their society. Thus, social exclusion is also a capability deprivation, which is the product of an interaction between structure and agency (Sen 2000). For example, in Chap. 5 of this book, the economic and housing reforms caused the low income and over-crowded living environment of the local residents in Xiguan. The residents decided to live in a dilapidated and deprived environment so that they could find affordable housing and informal jobs in Xiguan. Both social structures and agencies contribute to the exclusion of the residents in Xiguan from accessing well-being. To resolve social exclusion, transport planners require the transformation of government policies in accordance with the concepts of social justice.

The fifth objective of this book is to incorporate the concepts of the self-organisation approach into travel behaviour analysis and transport planning. Conventional travel demand forecasting and modelling fail to measure the uncertainties of travel behaviour that are produced by complex human social systems. This book suggests that travel behaviour is the outcome of a complex self-organisation process. That is, transport systems and the built environment are both the medium and the product of travel behaviours (Giddens 1984). The incorporation of the self-organisation process into conventional transport-planning models can help transport planners better understand the trends of future travel behaviours, as forecasting demands and providing solutions to transport problems seem to be difficult tasks for conventional urban transport planning models.

1.9 Methodology

The research framework of this book relies on the transport-related social exclusion model that has been discussed in the previous sections. According to the stages in the model, a combination of quantitative and qualitative data collection methods were used in the surveys that were conducted in Guangzhou and Shanghai. Since the complex travel behaviour of low-income workers cannot be effectively measured by surveys that focus on structural factors alone, research methods such questionnaires and in-depth interviews also asked about the attitudes and actions of the low-income workers in interacting with structural factors (Bohte et al. 2009). To consolidate the findings in qualitative surveys, this research also conducted commuting surveys on the work trips of low-income workers. In fact, the simple statistical analysis of work trips helps to summarise and determine the relationships among variables relevant to the complex travel behaviours of low-income workers in deprived neighbourhoods.

Many of the findings of the studies in this book are based on qualitative surveys. However, the adoption of qualitative surveys has to contend with the trade-off between the weak validity and reliability of the qualitative measures; for example, individual decisions are not directly observable, making it difficult to validate personal decision and attitude measures. For example, in the interviews, the author found that some respondents tended to answer 'yes' or agree with some items, while others intentionally distorted answers when they felt that the questions were too personal.

The qualitative survey in this book aimed to explore the influence of individual decisions and actions on commuting. The variables of structures and individual actions were operationalised in the measurable variables. The qualitative surveys were implemented in the case studies by means of in-depth interviews with low-income workers in the deprived neighbourhoods (such as Xiguan and the study urban village in Guangzhou) to explore their experiences and attitudes on the accessibility problems that are due to the continual interaction between structures and individual actions. The survey findings will help us to understand the impacts of the self-organisation process and the interaction between structure and agency on low-income workers' travel behaviours, social exclusion and accessibility problems.

To improve the validity and reliability of this research, quantitative measures, such as regression models, were used to measure the new behaviours that were produced by the self-organisation process in the case studies in Guangzhou. The results of the studies in Chaps. 4–6 of this book indicate that many transport-related social exclusion problems were related to the travel distance in work trips, suggesting that journey distance to work might be an appropriate measure of job access. Based on the concept of the space-time prism approach, this book will use the potential path area (PPA) to predict the accessibility level of an individual person. PPA is the spatial extent to which a person can overcome space-time constraints in order to participate in his desired activities. The shorter the distance a person commutes, the smaller the size of the PPA he or she experiences, the lower the numbers of desired activities he or she can participate in and thus the lower the level of accessibility he or she faces (Kim and Kwan 2003; Hägerstrand 1970). Lucas (2012) suggests that many low-income workers do not themselves see a need or necessarily want to travel outside their immediate local area (e.g., rural migrants in urban villages). This means that they cannot reach many important basic needs, such as medical treatments and recreational and social activities. This phenomenon is regarded as 'low travel horizons', and it is suggested that over time, social exclusion could act to constrain people's travel expectations.

In Chap. 6 of this book, we will see that transport studies have failed to identify the root cause of the ridership that is insufficient to support cost-effective rail transit services in Shanghai (Shen et al. 2016; Pan et al. 2011). Based on the investigation approach proposed by Schuetz et al. (2018), three on-site surveys were conducted to test the compatibility between the built-environment around the rail stations in the suburbs of Shanghai and the TOD design guidelines (Cervero et al. 2013). This book examines whether the declining rail ridership is due to problems of the development of the built environment or due to the problems that come from the self-organisation process. The author visited three rail stations in the suburbs and checked whether the population density and resettlement development density around the stations matched by the TOD design guidelines. The results of the on-site case studies indicate that TOD-friendly development alone was sufficient to develop and provide adequate travel demand for cost-effective rail services in the suburban areas of Shanghai. There must be other factors that were more important than TOD-friendly development in influencing the declining ridership of rail services. The self-organisation approach helped to nail down the cause and effect of the insufficient rail ridership.

1.10 Principal Hypotheses of this Book

Based on the concept of the self-organisation process, several case studies will be examined to investigate the interactions between structural factors and the individual actions of low-income workers in deprived urban neighbourhoods in Guangzhou and Shanghai. The case studies will follow three major hypotheses to operationalise the structural factors and individual actions in the self-organisation process, to determine the new travel behaviours that result in transport-related social exclusion, and to analyse the causes of accessibility problems and social exclusion. Furthermore, the hypotheses will guide this study to propose appropriate policy responses for Chinese government decision-makers, who want to build a just society, by minimising the social exclusion and accessibility problems of poor workers. The three principal hypotheses are as follows:

1. The self-organisation process, which is taking place in a study urban village, Qingfengcun, produces new structural properties, accessibility and social exclusion problems;
2. The social exclusion process, which is taking place in Guangzhou's historic inner city, Xiguan, produces new behaviours, poverty, accessibility inequity and capability deprivation; and
3. The process of the suburbanisation of poverty in Shanghai has produced new travel behaviours, social exclusion, accessibility problems and the declining ridership of rail services.

1.10.1 Elaboration of the Research Hypotheses

The first hypothesis is used to test whether the self-organisation process, which is taking place in the study urban village, Qingfengcun, produces social exclusion and accessibility problems that influence low-income workers. The hypothesis is

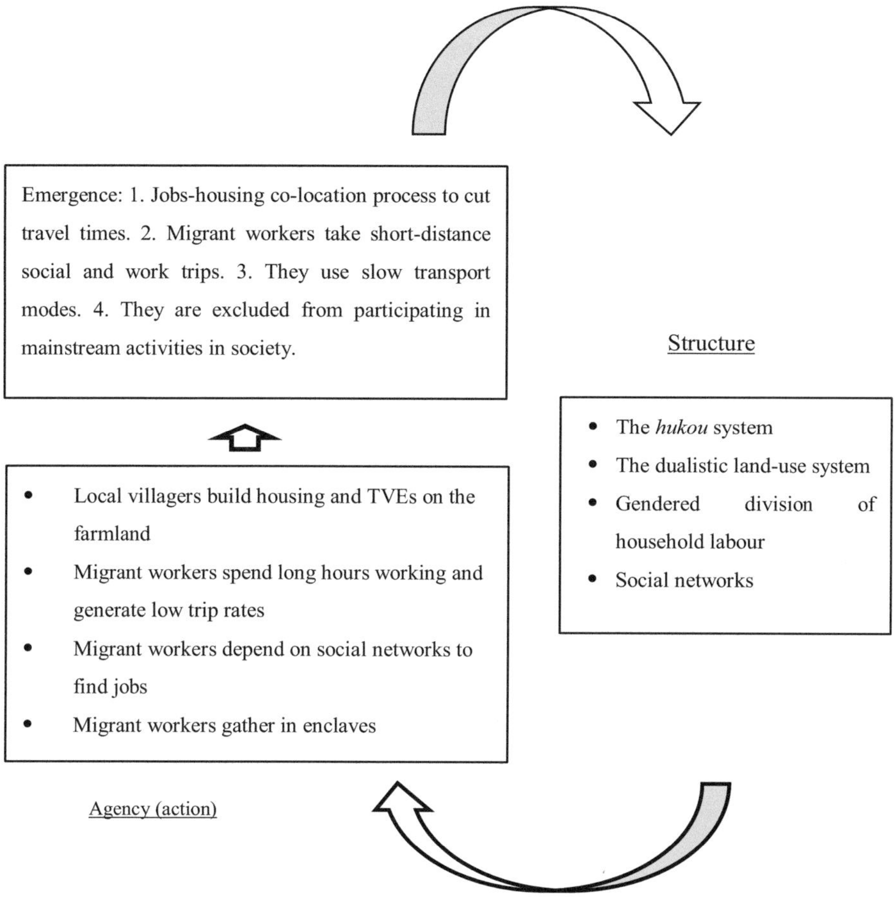

Fig. 1.3 Self-organisation process in the study village, Guangzhou

investigated in Chap. 4 of this book. It is hypothesised that the rules and social norms that shape the actions of migrants in the study urban village are both the medium and the product of the migrants' actions (Giddens 1984). Rural migrants and local villagers use their decisions and actions to transform rules and social norms through the reproduction of routine practices, and this process then produces emergence or new travel behaviours that lead to rural migrants' and local villagers' commuting problems and social exclusion. Thus, the self-organisation process can be regarded as a social exclusion process. The structures include the *hukou* system, the dualistic land-use system, the gendered division of household labour and the social networks in the study urban village (Fig. 1.3).

The *hukou* system restricts migrant workers from accessing formal job markets and social welfare in urban centres. They have to find jobs in the inner suburbs of Guangzhou. The dualistic land-use system allows village collectives to use a share of their farmland to develop TVEs and to build affordable housing on their own housing plots in the study urban village. The social norms (e.g., the gendered division of household labour) allocate more household responsibilities to working migrant mothers and restrict them from seeking distant job opportunities. The social networks contribute to the concentration of migrants, who are seeking to facilitate their job searches, in urban villages in Guangzhou (Zhu 2016; Zhao and Howden-Chapman 2010) (Fig. 1.3).

The individual actions of rural migrants include migrants producing high residential mobility to live near job opportunities in the study urban village, and their actions contribute to the formation of a jobs-housing co-location process that cuts travel times. The rural migrants spend long hours working in order to earn adequate income. They gather in the study urban village and depend on social networks to find jobs and affordable housing. They seldom take long-distance trips to contact locals and find recreational activities in the city centre, and they thus generate a lower level of trip frequency per day than the general population of the society.

Actions of migrants and local villagers transform the structures through the reproduction of practices to produce new travel behaviours. These new structural properties include the migrant women seeking jobs near their homes; migrants seek a high rate of residential mobility so that they can move swiftly to employment opportunities in urban villages, and they usually walk or ride bicycles to activities (Fig. 1.3).

In addition, the self-organisation process in the study urban village has produced new behaviours that lead to social exclusion problems. For example, migrant workers travel short-distance trips in their routine activities and contribute to the exclusion of migrants from participation in mainstream activities in Guangzhou. In fact, this hypothesis is used to show that the self-organisation approach can be used as a tool to investigate and minimise social exclusion. For example, based on the survey results, this book can provide suggestions to policy-makers in China to transform several government policies to improve migrants' accessibility and reduce their social exclusion problems.

The second hypothesis is used to test whether social exclusion is a type of self-organisation process that is shaped by the cycle of structure and agency. Social exclusion obstructs low-income workers' freedom and right to achieve well-being. Additionally, such exclusion is itself a type of self-organisation process. In the inner city study in Xiguan, social exclusion produces new behaviours, poverty, accessibility inequity and capability deprivation. It is hypothesised that the rules and social norms that shape the actions of the residents in the inner city study in Xiguan are both the medium and the product of the residents' actions (Giddens 1984). The local residents and migrants use their decisions and actions to transform rules and social norms through the reproduction of routine practices, and this process then produces emergence, which includes new behaviours that cause poverty, accessibility inequity and the capability deprivation of low-income residents. This hypothesis is investigated in Chap. 5 of this book, and the self-organisation process develops in the form of a social exclusion process.

In the 1930s, Xiguan was a popular neighbourhood for rich traders and was famous for its highly established commercial and cultural development. After 1949, many rich residents left Xiguan, the large houses were allocated to low-income workers in various work-units, and now, it has transformed into a deprived neighbourhood (Yuan and Wu 2014; Yuan et al. 2011).

In the social exclusion process, there have been several structural factors that cause the commuting problems and capability deprivation of residents. First, economic and housing reforms have caused local residents to live in an overcrowded and deprived environment. Second, because the streets are too narrow for transit buses, residents always experience conflicts between traffic and pedestrians during peak hours. Third, the privately funded and property-led redevelopment approach was unsuccessfully implemented in Xiguan due to low profits, and the Guangzhou government and residents want to preserve the traditional architecture of the large houses (Yuan et al. 2011).

Fourth, most of the local residents are redundant SOE workers; they can find only informal retail jobs in the neighbourhood due to the government revitalisation policy. Many old houses and streets have been converted into tourist walking streets. Finally, rural migrant workers face social discrimination when they compete with local residents over employment opportunities in the retail markets. In addition, rural workers are excluded from accessing formal job markets due to the *hukou* system (Fig. 1.4).

The actions of these migrants to transform structures have included two aspects. First, migrants have developed social networks to find jobs. Many of them are concentrated in the western side of Xiguan to find jobs and have gradually transformed the neighbourhood into a migrant enclave. Second, most migrants find work in informal retail positions in Xiguan, and they use slow transport modes, such as buses and bicycles, to reach their places of employment (Fig. 1.4).

There are several actions taken by local residents. First, the local residents pressure the local government on urban redevelopment to improve their indoor amenities (e.g., gas, shower and toilet) and to redevelop dangerous old buildings. Second, in the face of the inadequate housing supply, local residents subdivide their houses into more rooms and live in overcrowded housing environments (Tan and Altrock 2016; Yuan and Wu 2014; Yuan et al. 2011). Third, they make use of the revitalisation policy to find informal employment opportunities in Xiguan. Fourth, the women choose to find employment near their homes so that they can cut travel times to reach their jobs and have more time for childcare. Fifth, the local *hukou* residents, who own their flats, tend to have longer travel times to reach work than migrant workers do. Finally, the local residents have become accustomed to riding bicycles to work and non-work activities since the *danwei* era (Fig. 1.4).

The interactions between the actions of the residents produce new travel behaviours, which include commuting shorter distances to work, they use slower transport modes to reach activities, and they seldom take long-distance trips to reach social and recreational activities that are outside their own neighbourhood. They are excluded from participating in the mainstream activities of society (Fig. 1.4).

The third hypothesis is used to test whether the suburbanisation of the poverty process that takes place in Shanghai produces new behaviours that lead to social exclusion and low demand for rail transit services. The suburbanisation of the

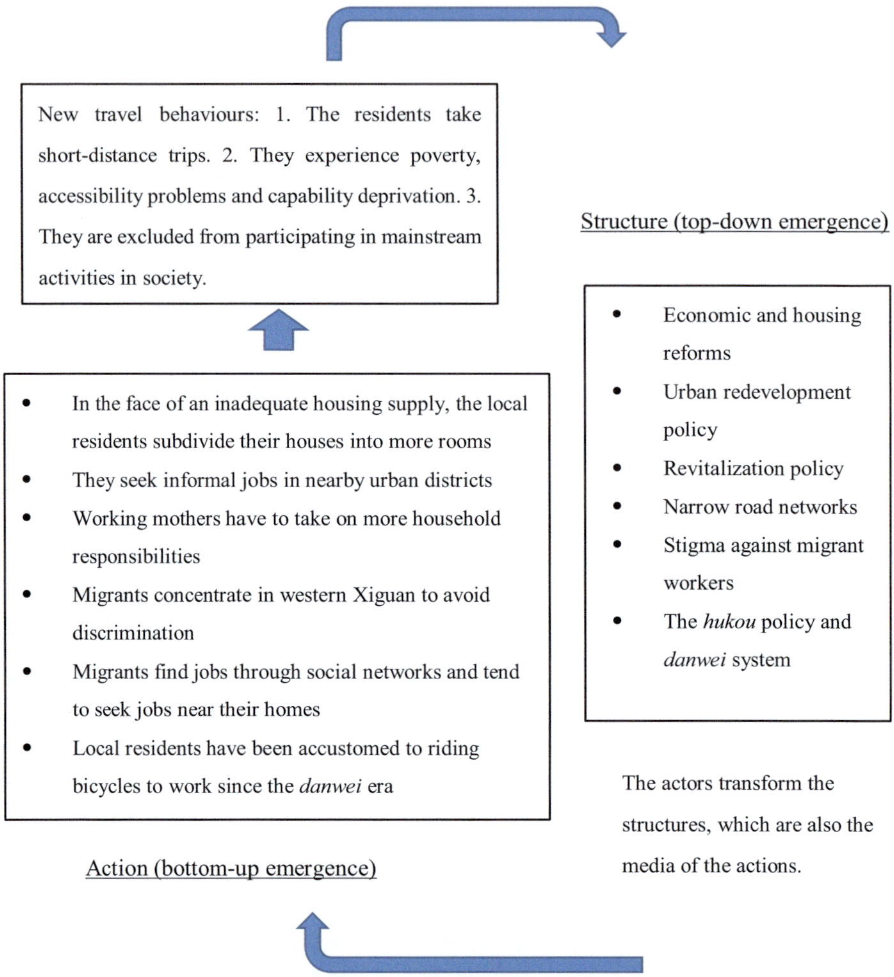

Fig. 1.4 Social exclusion process in Xiguan, Guangzhou

poverty process is a type of social exclusion process in which the mechanism of the process is structuration theory. This book hypothesises that interactions between government policies and individual actions in the process of suburbanisation in Shanghai develop an overarching trend of the suburbanisation of poverty towards the inner suburbs and surrounding regions (Giddens 1984). The process produces new structural properties, including a concentration of poverty in the inner suburbs, accessibility problems to places of employment and the capability deprivation of local residents to achieve well-being. The displaced residents and migrants use their decisions and actions to transform government policies and social norms through the reproduction of routine practices. The emergence and new behaviours also result in the trips of shorter distances taken by the displaced residents, which is the root cause of the declining ridership of the rail services in suburban areas. The hypothesis is tested and investigated in Chap. 6 of this book.

The structural factors in the process of the suburbanisation of poverty are as follows: First, the district governments in the suburban areas of Shanghai are evaluated on their achievements in suburbanising industries and populations; thus, they play a dominant role in building industrial zones and affordable housing for relocated residents during the process of suburbanisation. The government began acquiring farmland for building industrial zones, and this policy led to urban sprawl (Zhang et al. 2018; Liu et al. 2018a, b; You 2016). The government built large-scale rail transit networks to connect the new towns and the central city (Pan et al. 2017). The effectiveness and efficiency of the rail transport services were constrained by the urban sprawl and the new travel behaviours of the displaced residents.

Second, in the face of the *hukou* system, rural migrants failed to find jobs in urban areas; they chose to follow industrial jobs and move to the inner suburbs, where they developed migrant enclaves near large-scale industrial and residential zones. Third, migrants continue to make use of social networks to find jobs, and thus, their social activities and job activities are spatially restricted to the migrant enclaves. In 2016, rural migrants accounted for approximately 40.5% of the total population in Shanghai, and their travel behaviour and housing locations directly influence the performance of rail transit services in suburban areas (Shanghai Statistics Bureau 2017) (Fig. 1.5).

Fourth, during the process of suburbanisation, in order to generate revenue to finance its transport networks to link the new towns and the city centre, the municipal government carried out rapid and privately funded urban redevelopment and land finance policies. The property-lead redevelopment and land finance policies were driven by diverse motivations of the government, such as transferring land-use functions to business and commercial purposes and maximising benefits from land sales to finance the development of infrastructure. The policy also resettled a large number of low-income local residents from the inner city to distant new towns (Tian et al. 2017a, b, c; He and Wu 2007). The relocated residents were involuntarily moved to the suburban areas because the compensation they received from the demolition of their old inner city homes was not enough to buy new apartments in the inner city areas (Guan and Xu 2018; Chang and Chen 2013; Cervero and Day 2008) (Fig. 1.5).

The individual decisions and actions of low-income workers include rural migrants having chosen to live close to informal industrial employment opportunities in the inner suburbs. Some of them make use of strong social networks to develop enclaves and borrow start-up cash for developing small business activities, which are dispersed around the suburban areas. Most of the displaced low-income local workers were involuntary moved to the suburban areas. Because they are low skilled and have low levels of education, most of the middle-aged workers have failed to find new jobs in suburban areas and have chosen to retire early. Younger individuals with higher educational achievements have chosen to travel long distances and find high-paying jobs in the city centre (Guan and Xu 2018; Yuan and Lin 2013).

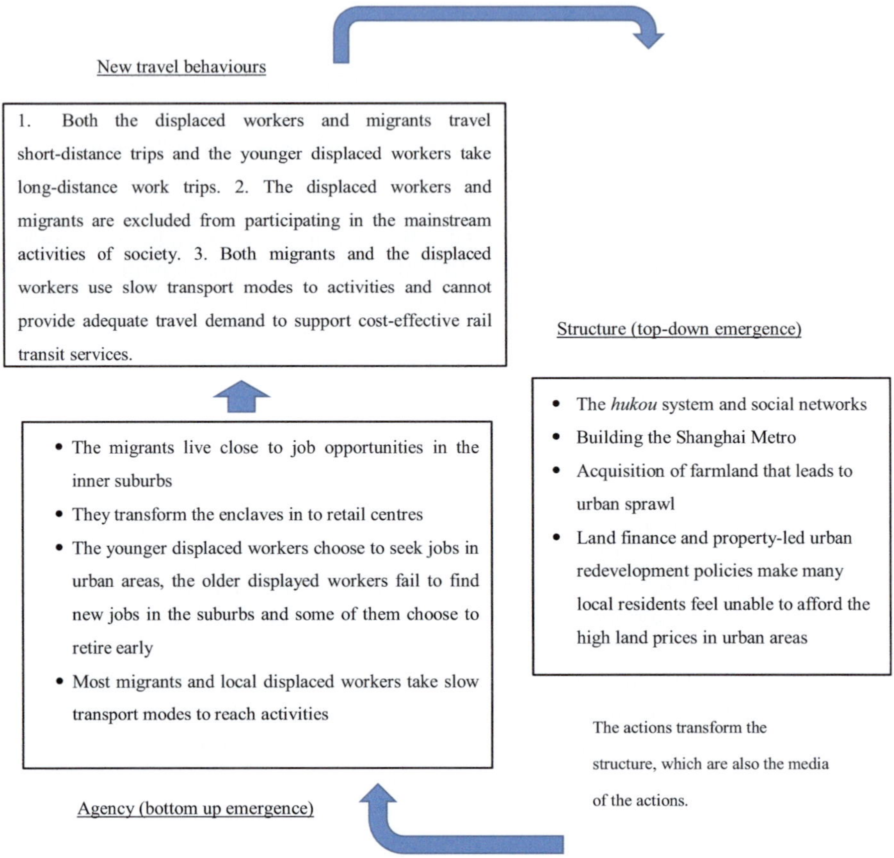

Fig. 1.5 Process of the suburbanisation of poverty in Shanghai produces social exclusion, accessibility problems and declining rail ridership

The self-organisation process has produced new travel behaviours, which include younger displaced residents travelling long distances from their homes to their places of employment in urban areas. A high number of middle-aged and unskilled displaced residents have failed to find new jobs, and they take short-distance non-work trips and use slow transport modes to reach activities. Migrant workers live close to job opportunities, and they travel short distances on social and work trips. They choose slow transport modes to reach activities. The displaced workers and migrants are excluded from participating in the mainstream activities of their society.

In addition, the author argues that the built environment of the rail stations in the suburban areas is not the main cause of the declining rail transit ridership in these areas of Shanghai. The process of suburbanisation of poverty in Shanghai, which produces the displaced residents' short-distance trips, causes the declining demand for rail services. Route tests and on-site surveys were conducted to test this argument (Fig. 1.5).

1.11 The Organisation of the Book

The book is structured into seven chapters. Chapter 2 reviews the theoretical framework of the self-organisation process and explains the concepts of the structuration theory, which is used as the mechanism of the self-organisation approach. The chapter also examines the duality of the structure identified as a typical self-organisation process that produces new structural properties and travel behaviours. This book finds that the self-organisation process can be converted into a social exclusion process that produces new accessibility problems and social exclusion in social systems. In the self-organisation process, the structure is identified as the root cause of accessibility problems and social exclusion. The book suggests that there are limitations in using policies alone to investigate travel behaviour and social exclusion. This is because the structure (e.g., transport rules) is the medium and outcome of travel practices (Giddens 1981, 1984). New travel behaviours are the product of the self-organisation process, which is the interaction between structure and travel practice. Hence, the self-organisation process is identified as an effective research tool that incorporates structure and agency to investigate travel behaviours.

Chapter 3 reviews the concepts of accessibility approaches and the application of capability and self-organisation approaches to accessibility investigation. The chapter also investigates the important role of individual decisions and actions in the self-organisation process to achieve basic needs. Using the travel-time ratio concept, the chapter also analyses the influence of individual decisions on the allocation of time intervals to achieve basic needs. Furthermore, the concepts of the space-time prism and potential path area are introduced to measure the accessibility of low-income workers in the self-organisation process. Travellers follow rules to reproduce daily travel practices and transform the rules in routine commuting. The process produces new properties that give rise to space-time accessibility constraints and negatively influence their travel distance, mode choice and trip frequency when participating in activities.

Since unfair government policies are found to be the root cause of accessibility problems and social exclusion, Chap. 3 introduces the theories of justice (e.g., the utilitarian, Rawlsian and capability approaches) to help government policy-makers identify unjust policies and adopt social justice theories to deliver equitable policies for minimising social exclusion and improving accessibility.

Chapter 4 presents an investigation on the social exclusion of migrants in accessing non-work activities in an urban village in Guangzhou. Rural migrants feel unable to find jobs in the urban areas of Guangzhou because of the *hukou* system. In their effort to find jobs, rural migrants become concentrated in the study village and use their social networks to find affordable housing and informal manufacturing jobs. In addition, the dualistic land-use system allows village collectives to build TVEs on their farmland and multi-storey buildings on housing plots. The interactions among structural factors and the actions of migrants produce new travel behaviours, which include migrant workers taking short-distance trips. They use slow transport modes to reach their places of employment and show a low level of trip rates for non-work activities. The new travel behaviours contribute to the exclusion of migrants from participating in the mainstream activities of their society, as they encounter accessibility problems in making social contacts with local residents and in reaching recreational and medical facilities.

Chapter 5 presents a study on the social exclusion experience of low-income workers in Xiguan, an historic inner city in Guangzhou. The study investigates how the self-organisation process excludes residents from accessing job opportunities and decent homes and living environments, e.g., how market and housing reforms interact with the actions of the residents. The local government tries to revitalise the economy of the inner city by developing retail markets for tourists and redeveloping old houses. Migrant residents face social discrimination, and the *hukou* system excludes them from accessing formal employment opportunities. To avoid the impacts of stigmatisation, they concentrate on the western side of Xiguan in areas such as the retail streets for tourists and the wholesale clothing market to find informal employment. As a result, the self-organisation process produces new travel behaviours, which include low-income workers' short-distance travel patterns, use of slow transport modes to access activities, and low trip rates for accessing social and recreational activities. As a result, the process excludes low-income residents from participating in the mainstream activities of their society, such as finding formal jobs, making distant social contacts, and accessing recreational facilities.

Chapter 6 investigates the suburbanisation process (or a social exclusion process) in Shanghai that has produced the concentration of poverty in the inner suburbs of Shanghai. The process developed in the form of the self-organisation process. It produced the commuting and social exclusion problems of the low-income workers, resulting in the declining ridership of rail transit services. Government policies, such as land financing, property-led urban redevelopment, farmland acquisition, industrial zone sprawl, the *hukou* system and the construction of rail transit systems, have contributed to the suburbanisation of poverty and the exclusion of displaced residents from participating in the mainstream activities of society. Many displaced middle-aged residents, most of them workers with low levels of education, have failed to find new jobs in the suburban areas and have chosen to retire early; thus, these residents take short-distance trips. The rural migrants, who have chosen to gather in migrant enclaves, depend on social networks to find jobs and borrow start-up cash for developing small retail businesses live near informal jobs in the suburbs of Shanghai. The social exclusion process has produced new travel behaviours, which include these residents' concentration in the suburbs, taking of short-distance trips, use of slow transport modes and low trip rates for accessing non-work activities.

In addition, by using on-site investigations and route tests, this chapter investigates whether the built environment, which was not built using TOD guidelines, is the root cause of the declining rail ridership in the suburban areas of Shanghai. The findings of the investigations confirm that the built environment alone is not the root cause, and this book argues that the social exclusion process that produces the short-distance trips of low-income workers contributes to the declining ridership of the rail services.

Chapter 7 aims to conclude the roles of the self-organisation process in transport planning and social exclusion by introducing a new transport-related social exclusion model, which is also a self-organisation process. The model introduces procedures in transport planning to identify the impacts of structural factors on the commuting and social exclusion of low-income workers, the influence of individual actions on commuting problems and social exclusion, and the new behaviours that contribute to the exclusion of low-income workers. The model also introduces steps to adopt social justice theories and a collaborative approach to improve social inclusion and commuting. In addition, based on the concept of neighbourhood effects, the travel behaviours of the low-income workers in the deprived neighbourhoods should be different from their travel behaviours in other urban neighbourhoods in a city. Hence, conventional transport planning models, which measure aggregated travel demand, may obscure the travel behaviours in deprived neighbourhoods. This chapter introduces steps to incorporate the findings of the transport-related social exclusion model into the conventional four-step transport-planning model to improve the accuracy of travel demand forecasts and improve the accessibility of low-income workers.

References

Andersson M, Brundell-Freij K, Eliasson J (2017) Validation of aggregate reference forecasts for passenger transport. Transp Res A 96:101–118

Batty M (2012) Urban regeneration as self-organisation. Architectural Design-Special Issue: London Regeneration 82(1):54–59

Bliemer MCJ, Raadsen MPH, Brederode LJN, Bell MGH, Wismans LJJ, Smith MJ (2017) Genetics of traffic assignment models for strategic transport planning. Transp Rev 37(1):56–78

Blumenberg E (2004) En-gendering effective planning: spatial mismatch, low-income women, and transportation policy. J Am Plan Assoc 70(3):269–281

Bohte W, Maat K, Van Wee B (2009) Measuring attitudes in research on residential self-selection and travel behaviour: a review of theories and empirical research. Transp Rev 29(3):325–357

Boisjoly G, El-Geneidy AM (2017) The insider: a planners' perspective on accessibility. J Transp Geogr 64:33–43

Cao M, Hickman R (2019) Understanding travel and differential capabilities and functionings in Beijing. Transp Policy 83:46–56

Cervero R, Day J (2008) Suburbanization and transit-oriented development in China. Transp Policy 15:315–323

Cervero R, Iuchi K, Suzuki H (2013) Transforming cities with transit: transit and land-use integration for sustainable urban development, urban development series. World Bank

Chang Y, Chen J (2013) Public housing in Shanghai: a tool with multiple purposes. Housing Finance International, Winter, pp 27–34

China Labour Bulletin (2017) Migrant workers and their children, China. https://clb.org.hk/content/migrant-workers-and-their-children. Accessed 19 Nov 2018

de Jong G, Daly A, Pieters M, Miller S, Plasmeijer R, Hofman F (2007) Uncertainty in traffic forecasts: literature review and new results for the Netherlands. Transportation 34:375–395

Elvy J (2014) Public participation in transport planning amongst the socially excluded: an analysis of 3rd generation local transport plans. Case Stud Transp Policy 2:41–49

Flyvbjerg B, Holm MKS, Buhl SL (2005) How (in)accurate are demand forecasts in public works projects? J Am Plan Assoc 71(2):131–146

Fuchs C (2003) Structuration theory and self-organization. Syst Pract Action Res 16(2):133–167

Fuchs C, Hofkirchner W (2005) The dialectic of bottom-up and top-down emergence in social systems. TripleC i(i):28–50

Gershenso C (2018) Self-organizing urban transportation systems, Universidad Nacional Aut´ onoma de M´exico. https://arxiv.org/pdf/0912.1588.pdf. Accessed 4 Nov 2018

Giddens A (1981) A contemporary critique of historical materialism. Vol. 1. Power, property and the state. Macmillan, London

Giddens A (1984) The constitution of society. Polity Press, Cambridge

Giuliano G, Small KA (1993) Is the journey to work explained by urban spatial structure? Urban Stud 30(9):1485–1500

Guan J, Xu C (2018) Are relocatees different from others? Relocatee's travel mode choice and travel equity analysis in large-scale residential areas on the periphery of megacity Shanghai, China. Transp Res A 111:162–173

Hägerstrand T (1970) What about people in regional science? Reg Sci Assoc Pap 24(1):7–21

He S, Wu F (2007) Socio-spatial impacts of property-led redevelopment on China's urban neighbourhoods. Cities 24(3):194–208

Healey P (1997) Collaborative planning: shaping places in fragmented societies. UBC Press, Vancouver

Hong Kong Transport Department (2014) Travel characteristics survey 2011: final report, Hong Kong SAR Government

Hong Kong Transport Department (2018) 2018 Annual transport digest, Hong Kong SAR Government

Kim HM, Kwan MP (2003) Space-time accessibility measures: a geo-computational algorithm with a focus on the feasible opportunity set and possible activity duration. J Geogr Syst 5(1):71–91

Krugman PR (1996) What economists can learn from evolutionary theorists, (transcript of speech given to the European Association for Evolutionary Political Economy, Rome). http://web.mit.edu.eproxy1.lib.hku.hk/krugman/www/evolute.html

Liu CY, Ye L, Feng B (2018a) Migrant entrepreneurship in China: entrepreneurial transition and firm performance. Small Bus Econ 52:1–16

Liu Y, Fan P, Yue W, Song Y (2018b) Impacts of land finance on urban sprawl in China: the case of Chongqing. Land Use Policy 72:420–432

Lucas K (2012) Transport and social exclusion: where are we now? Transp Policy 20:105–113

McNally MG (2000) Four step model. Institute of Transportation Studies, University of California, Irvine, CA

Meyer D, Miller EJ (2001) Urban transport planning: a decision-oriented approach. McGraw-Hill, New York

Morrison K (2005) Structuration theory, habitus and complexity theory: elective affinities or old wine in new bottles? Br J Sociol Educ 26(3):311–326

Neto RS, Duarte G, Páez A (2015) Gender and commuting time in São Paulo metropolitan region. Urban Stud 52(2):298–313

Pan H, Shen Q, Liu C (2011) Transit-oriented development at the urban periphery: insights from a case study in Shanghai, China. Transp Res Rec 2245:95–102

Pan HX, Li J, Shen Q, Shi C (2017) What determines rail transit passenger volume? Implications for transit oriented development planning. Transp Res D 57:52–63

Pei K, Fan D (2010) Passenger flow forecast for urban rail transit based on extended four-step model in Wuxi, China. J Express Rail Traffic 23(5):57–61. (in Chinese)
Pereira RHM, Schwanen T, Banister D (2017) Distributive justice and equity in transportation. Transp Rev 37(2):170–191
Portugali J (2000) Self-organization and the city. Springer, Heidelberg
Portugali J (2006) Complexity theory as a link between space and place. Environ Plan A 38:647–664
Preston J, Rajé F (2007) Accessibility, mobility and transport-related social exclusion. J Transp Geogr 15:151–160
Pyrialakou VD, Gkritza K, Fricker JD (2016) Accessibility, mobility, and realized travel behavior: assessing transport disadvantage from a policy perspective. J Transp Geogr 51:252–269
Rawls J (1971) A theory of justice. Oxford University Press, Oxford
Ryan J, Wretstrand A, Schmidt SM (2019) Disparities in mobility among older people: findings from a capability-based travel survey. Transp Policy 79:177–192
Schuetz J, Giuliano G, Shin EJ (2018) Does zoning help or hinder transit-oriented (re)development? Urban Stud 55(8):1672–1689
Schwanen T, Lucas K, Akyelken N, Solsona DC, Carrasco JA, Neutens T (2015) Rethinking the links between social exclusion and transport disadvantage through the lens of social capital. Transp Res A 74:123–135
Sen A (1980) Equality of what? Choice, welfare and measurement (Cambridge, MA: Harvard University Press), 1997 [Originally published in The Tanner Lectures on Human Values, Vol. I, by the University of Utah Press and Cambridge University Press, 1980]
Sen A (1981) Poverty and famines: an essay on entitlement and deprivation. Clarendon Press, Oxford
Sen A (1984) Resources, values and development. Oxford, Basil Blackwell
Sen A (1992) Inequality reexamined. Oxford University Press, Oxford
Sen A (2000) Social exclusion: concept, application and scrutiny. Social development papers no 1. Office of Environment and Social Development, Asian Development Bank, Manila
Shanghai Statistics Bureau (2017) Shanghai statistical year book 2017, Shanghai Municipal Government (in Chinese)
Shen J (2017) Stuck in the suburbs? Socio-spatial exclusion of migrants in Shanghai. Cities 60:428–435
Shen Q, Chen P, Pan H (2016) Factors affecting car ownership and mode choice in rail transit-supported suburbs of a large Chinese city. Transp Res A Policy Pract 94:31–44
Smith A (1937) An inquiry into the nature and causes of the wealth of nations [1776]. In: Cannan E (ed) . Random House, New York
Stacey R (1996) Complexity and creativity in organizations. Berrett-Koehler, San Francisco
Tan XH, Altrock U (2016) Struggling for an adaptive strategy? Discourse analysis of urban regeneration processes – a case study of Enning road in Guangzhou City. Habitat Int 56:245–257
Tian L, Ge B, Li Y (2017a) Impacts of state-led and bottom-up urbanization on land use change in the peri-urban areas of Shanghai: planned growth or uncontrolled sprawl? Cities 60:476–486
Tian L, Guo X, Yin W (2017b) From urban sprawl to land consolidation in suburban Shanghai under the backdrop of increasing versus decreasing balance policy: a perspective of property rights transfer. Urban Stud 54(4):878–896
Tian L, Li Y, Shao L, Zhang Y (2017c) Measuring spatial-temporal characteristics of city expansion and its driving forces in Shanghai from 1990 to 2015. Chin Geogr Sci 27(6):875–890
United Nations (2016) Identifying social inclusion and exclusion. In: Leaving no one behind: the imperative of inclusive development, report on the social situation 2016. Department of Economic and Social Affairs, New York, pp 17–28
Van Wee B, Geurs K (2011) Discussing equity and social exclusion in accessibility evaluations. Journal of Emerging Technologies and Innovative Research 11(4):350–367

Wang D, Gan D, Zhu C, Lu X (2012) A research on the relationships between suburban spatial planning and railway traffic planning in Shanghai. J City Plan 1:17–22. (in Chinese)
Wee BV, Roeser S (2013) Ethical theories and the cost–benefit analysis-based *Ex Ante* evaluation of transport policies and plans. Transp Rev 33(6):743–760
Xia JC, Nesbitt J, Daley R, Najnin A, Litman T, Tiwari SP (2016) A multi-dimensional view of transport-related social exclusion: a comparative study of greater Perth and Sydney. Transp Res A 94:205–221
You H (2016) Quantifying megacity growth in response to economic transition: a case of Shanghai, China. Habitat Int 53:115–122
Yuan J, Lin H (2013) An empirical study of commuting characteristics in rural areas. Procedia Soc Behav Sci 96:114–122
Yuan Y, Wu FL (2014) The development of the index of multiple deprivations from small-area population census in the city of Guangzhou. PRC Habitat Int 41:142–149
Yuan Y, Wu FL, Xu XQ (2011) Multiple deprivations in transitional Chinese cities: a case study of Guangzhou. Urban Aff Rev 47(5):721–736
Zhang SH, De Roo G (2016) Interdependency of self-organisation and planning: evidence from Nanluoguxiang Beijing. Town Plan Rev 87(3):253–274
Zhang L, Yue W, Liu Y, Fan P, Wei YD (2018) Suburban industrial land development in transitional China: spatial restructuring and determinants. Cities 78:96–107
Zhao P, Howden-Chapman P (2010) Social inequalities in mobility: the impact of the *hukou* system on migrants' job accessibility and commuting costs in Beijing. IDPR 32(3–4):363–384
Zhou M, Wang D (2018) Investigating inter-generational changes in activity-travel behavior: a disaggregate approach. Transportation. https://doi-org.eproxy.lib.hku.hk/10.1007/s11116-018-9863-x. Accessed 4 Nov 2018
Zhou SH, Deng LF, Kwan MP, Yan RG (2015) Social and spatial differentiation of high and low income groups' out-of-home activities in Guangzhou China. Cities 45:81–90
Zhu P (2016) Residential segregation and employment outcomes of rural migrant workers in China. Urban Stud 53(8):1635–1656
Zhu PY, Zhao SN, Wang LP, Yammahi SA (2017) Residential segregation and commuting patterns of migrant works in China. Transp Res D 52:586–599

Chapter 2
Self-Organisation, Travel Behaviour and Social Exclusion

2.1 Introduction

A number of studies have examined the built-environment, transport system and rules that affect travel behaviour, but few have considered how commuting patterns are influenced by the interactions between social structures (e.g., rules and resources) and individual actions. To fill this research gap, this chapter investigates the influence of the cycle of structures and actions on individual travel behaviours. Urban neighbourhoods in China are complex social systems. Self-organisation processes will occur when urban neighbourhoods face external interventions, such as urban redevelopment projects and the *hukou* system, which cause many residents in urban areas to take actions to cope with changes.

When a social system faces a state of disequilibrium (e.g., migrant workers fail to find jobs in urban China), a self-organisation process will take place, which will produce emergence or new structural properties in the internal process to find an optimum fit with the environment. Emergence is not well defined, but all emergent social systems have the following characteristics: the whole is more than the sum of the parts in the social system because new structural properties have developed. In urban transport planning, the structuration theory is used as the mechanism of self-organisation for urban transformation (Giddens 1981, 1984). Structuration theory states that structure is both the medium and the outcome of the reproduction of an agency.

Based on structuration theory, the self-organisation process is divided into two stages: bottom-up emergence and top-down emergence. Bottom-up emergence refers to agency; top-down emergence refers to structural factors (Fuchs and Hofkirchner 2005). Top-down emergence hinders or enables an individual in achieving his or her intended goals by setting rules (e.g., the *hukou* system) that limit his or her scope of action and providing resources (e.g., social networks) to facilitate his or her action. Hence, the structure is the medium of action (Giddens 1984).

J. C.-y. Lau, *Self-Organisation Shapes Travel Behaviours and Social Exclusion in Deprived Urban Neighbourhoods of China*,
https://doi.org/10.1007/978-981-15-2252-9_2

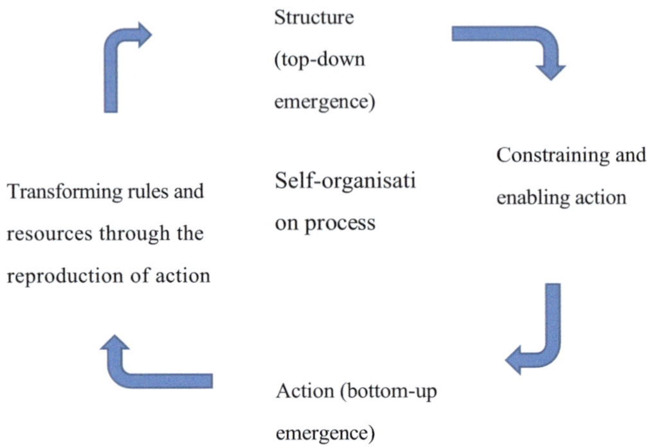

Fig. 2.1 Self-organisation process in social systems

At the same time, the actor in the social system transforms the structure through the production and reproduction of his or her routine actions. The self-organisation process produces new travel behaviours, such as travellers commuting short distances to reach work and using slow transport modes. The new behaviours gradually transform the structural factors, as even though the *hukou* system prevents rural migrants form finding jobs in the city centre of Guangzhou, they can find informal jobs in the inner suburbs. Thus, the structure is also the product of action (Giddens 1984). The self-organisation process will continuously evolve to find an optimum fit with the environment, as will occur when migrant workers can find employment and affordable housing and survive in the inner suburbs of Guangzhou (Fig. 2.1).

The self-organisation approach matches the requirements of social exclusion because both approaches involve multi-dimensional phenomena (e.g., people cannot participate in the mainstream activities of society due to resource deprivation, unjust policies, a lack of freedom and opportunities), and they both involve continual interactions between social policies and individual actions that produce new behaviour and social exclusion problems. For this reason, the self-organisation approach can be used as a tool to investigate transport-related social exclusion. The self-organisation process not only produces new travel behaviours (or emergent properties) of low-income workers but also uncovers the underlying characteristics of travel behaviours that have not been identified in existing transport studies.

2.2 Self-Organisation and Complex Travel Behaviours

One of the objectives of this book is to investigate the influence of self-organisation on the complex travel behaviour of low-income workers in urban China. Since complex travel behaviours are produced by a complex social system, this section will first investigate the characteristics of a complex social system. Then, it will introduce the relationships between a complex social system, the self-organisation process and new behaviours.

Human society is a complex and self-organising system (Batty 2009; Portugali 2006; Fuchs 2003). A complex system can be defined as a system that is open and complex. The system is 'open' in that the system exchanges matter, norms, and personal values with its environment. Such systems are so complex that it is difficult to predict and simulate the actions and interactions between their internal members or to follow how such systems give rise to new structural properties (Portugali 2006; Morrison 2005; Fuchs 2003). Some dilapidated inner-city neighbourhoods and migrant enclaves in urban China are open social systems that exhibit non-linearity and spontaneous interactions between individual actions and social rules.

Self-organisation is a property of complex social systems that produce complex new behaviours. Portugali states that:

> Self-organization is a property of systems that are open and complex. No one plans such systems, no one fully controls them and yet they have order, rules, and organization, and all these emerge spontaneously by means of self-organization. (Portugali 2017, p. 7)

In certain situations, external forces act on social systems and trigger internal and independent processes by which social systems spontaneously self-organise (Portugali 2017). For example, the gendered division of household labour in two-worker households allocates more household responsibilities to working mothers than to working fathers. The working mothers self-organise the activities that produce emergent properties or new travel behaviours, such as finding jobs that are in proximity to their homes. In this case, the social norms, e.g., the gendered division of household labour, are social rules, and households are complex social systems.

The self-organisation approach emphasizes a mechanism through which human communities self-organise their internal structure and continuously evolve to find an optimum fit with their environment (Portugali 2000; Heylighen 2008). In social science, the word 'evolve' refers to knowledgeable human actors who are dissatisfied with the environment of their social system and who transform the structure of the system through the production and reproduction of routine practices that result in new structural properties (Fuchs 2002, 2003). The self-organisation process involves spontaneous individual actions with no collective intention, and it is influenced by individual decisions, attitudes, and practices to reach intended goals. The process produces an emergent phenomenon (e.g., the new travel patterns of migrant workers in urban villages), resulting in a sudden transition of the social system to a state with qualitatively different structural properties (Kloch et al. 2011). For example, rural migrants who act autonomously with no collective intention can still produce meaningful changes in the housing and industrial production in urban villages in China.

In human social systems, sometimes, the self-organisation process takes place in one form of the social exclusion process in which unjust government policies interact with the agency of low-income workers to produce social exclusion problems. For example, the *hukou* system interacts with the actions of rural migrant workers and produces new structural properties, including the exclusion of migrants from formal labour markets in urban China. Government rules like those above can cause an imbalanced situation in the social system and reach a critical point that influences the quality of life or the survival of the actors in a social system. In such cases, bottom-up emergence takes place. The affected actors carry out a non-linear process of adjusting their behaviours to restore the balance of the system to achieve their intended goals, such as gaining employment. The process produces new structural properties at specific times, and in this study, the new properties include the new travel behaviours of low-income workers to reach their places of employment. In addition, the new travel behaviours have led to social exclusion problems, such as accessibility problems in reaching non-work activities (Rauws et al. 2016).

The above self-organisation process also happens in insect social systems, such as that of honeybees. When swarms of honeybees want to avoid danger, they move together, and sometimes, the swarm behaves as if it were a single giant animal. The individual honeybees follow a natural mechanism to interact with each other and form the different collective behaviours of the bee swarm, which are emergent phenomena and do not involve any central coordination (Heylighen 2001).

In addition, self-organisation appears in the free market system. Adam Smith (1937) proposed the idea of the invisible hand as a self-organisation mechanism for organising the interactions among market components and producing a state of equilibrium in the free market system. Recently, the concept of self-organisation has been adopted by regional and urban studies (Portugali 2000; Batty 2009; De Roo 2016; Zhang and De Roo 2016). It is important to understand the self-organisation process because the most effective way to manage cities is to discover their self-organisation mechanisms and then to work with those mechanisms to guide the process along a desirable route (Ball 2013).

Individual decisions and actions in the self-organisation process play an important role in producing new travel behaviours in terms of travel time, travel distance, mode choice, trip frequency and trip purpose. For example, Mote and Whitestone (2011) found that drivers, based on their individual decisions and actions, developed the spontaneous travel behaviour of slugging, an informal system of carpooling in the Washington, D.C. area, to interact with the traffic rules of high-occupancy vehicle (HOV) lanes in the early 1970s. Many single drivers picked up riders alongside the road (slugs) in order to meet the requirements for driving in the less congested HOV lanes.

In the above case, the drivers transformed the traffic rules through the production and reproduction of their routine driving practices (slugging) to improve their mobility: they picked up riders alongside the road to enable them to use the HOV lanes. There were several traffic lanes, and the drivers could choose to use either the HOV lanes or the other, more congested lanes. Therefore, it is important to provide a favourable environment so that individual drivers have the capability (e.g., the availability of freedom, choice and opportunities) to exercise their decisions in the course of executing their driving practices and to transform the traffic rules as part of the self-organisation process (Beyazit 2011; Sen 1999).

2.3 Mechanism of Self-Organisation Process: Structuration Theory

Both structuration theory and concepts of self-organisation stress that the reproduction of human social systems at the centre of their theories, such as systems, can be recursively self-producing. According to the structuration theory, the recursiveness of social life is an essential concept and can be shaped by the duality of structure. This means that the structural properties of social systems are both the medium and the outcome of the practices that constitute those systems (Morrison 2005; Fuchs 2003; Mingers 1995; Giddens 1981). Therefore, the mechanism of the self-organisation process in human social systems can be shaped by structuration theory (Giddens 1981, 1984).

Agency refers to an actor's capability of executing his or her individual decisions and actions to interact with the structure to achieve his or her choices. Social structure comprises the rules and resources that shape an actor's actions (Giddens 1981, 1984). The constitution of human social systems is based on the interaction between structure and agency, in structure is both the medium and the product of agency. Structures enable and constrain actors' practices but are also transformed by those practices. Structures are enacted by what Giddens calls 'knowledgeable' human actors (e.g., social norms, traditions and values are stored in individuals' memories, which constrains and enables their future practices; thus, actors know what they are doing and how to do it in their routine practices). In working to achieve their intended goals, human agents are capable of transforming structures through the reproduction of practices in the course of routine social interactions to produce new structural properties. This recursive process is called the duality of structure, which is defined as 'both the medium and the outcome of the practices which constitute social systems' (Giddens 1981, p. 27). Structure and agency are a duality; neither one can exist without the other.

Structures are also rules that create meaningful symbolic systems that provide ways for actors to perceive and interpret events (e.g., the *hukou* system denies rural populations access to formal employment in urban areas). The stores of knowledge result in communication among social actors (e.g., rural migrants). The structures thus influence the manner in which actors understand themselves and attach meaning to their surroundings.

The self-organisation process includes the individual decisions and actions of actors. For example, migrants make individual decisions and take individual actions to tackle the problems imposed by the *hukou* system. Their individual decisions and actions in urban villages include gathering in migrant enclaves, finding affordable housing and making use of social networks with their fellow villagers to seek out jobs. In addition, some of them make use of social capital to borrow start-up cash and develop small manufacturing production businesses. These migrants work long hours in industrial production enterprises in the enclaves. Migrant enclaves, or urban villages, are deprived of public services, such as educational, recreational, sports, healthcare, public transit, and postal service facilities. These facilities are

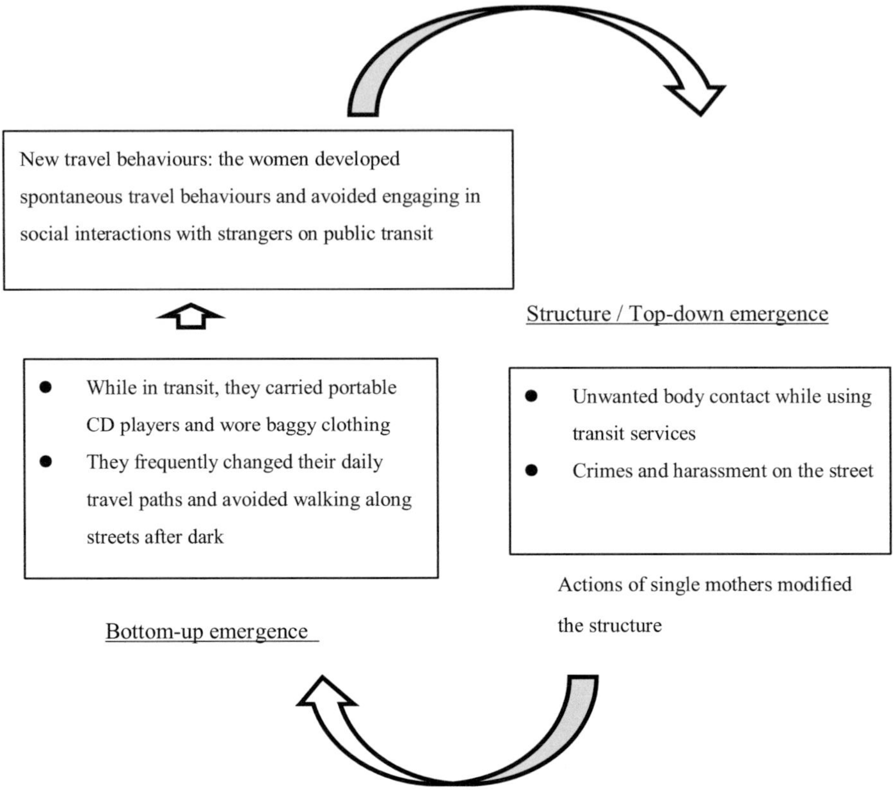

Fig. 2.2 Top-down and bottom-up emergence in a case study of single mothers

located far away, and migrants do not have time to take long-distance trips to meet these basic needs. As a result, the interactions between the structures and the actions of migrants exclude them from participating in the mainstream activities of society (Ouyang et al. 2017; Zhu et al. 2017; Jiading District Bureau of Statistics 2014).

The availability of freedom and choice for individual actions and decisions is an important variable for the quality of life of low-income workers. Individuals differ fundamentally in their capability to translate rules and resources into actions. The freedom and choice for individual actions depend on whether the government decision-makers in China, who want to build a just society in the country's urban areas, follow social justice concepts in the process of policy formulation (Sen 1992).

The structuration theory shapes the routine activities of actors. In transport investigation, the theory can be operationalised in terms of the two-stage self-organisation process, top-down emergence and bottom-up emergence (Fuchs 2002, 2003). A case study about the interaction between the social structure and individual actions of a group of low-income single mothers in San Francisco is used to explain the process (McQuoid and Dijst 2012) (Fig. 2.2).

The top-down emergence in this case is that the single mothers felt that their personal safety was at risk in their routine commuting practices. They experienced unwanted social and body contact while using public transit services. They feared walking along streets where they had little control over their exposure to harassment and crimes. The impetus of external social influences, such as harassment, crimes and unwanted social contact, is called top-down emergence. The bottom-up stage was that the single mothers made individual decisions and took individual actions to protect their personal safety. While on public transport, they carried portable CD players and iPods and wore baggy clothing to discourage unwanted social interaction. In addition, they frequently changed their routine travel paths and avoided walking on the street after dark. These new actions did not involve any central coordination and were the consequence of single mothers' independent decisions. The process produced new travel behaviours, which included the single mothers making spontaneous new travel paths that directly influenced the accessibility of necessary activities, and they avoided making social contact with strangers while using transit services.

2.4 Social Structure Is the Root Cause of Accessibility Problems

The mechanism of the self-organisation process is based on structuration theory. In addition, structure can be represented by transport policies in urban transport planning. Dimitriou (2013) suggests that social structure, constituted by such entities as the housing market, the labour market and transport planning, are the root causes of commuting problems. Dimitriou (2013) states that these causes are:

> Movements which are seen to be a function of land use and socio-economic changes of urban areas (Dimitriou 2013, p. 153)

Therefore, economic policies (e.g., municipal governments trying to achieve high GDP growth in urban China) are the main root causes of rapid suburbanisation, the spatial mismatch of jobs, and the housing and commuting problems of low-income workers in urban China.

For example, the housing and economic reforms in China prevented the redundant SOE workers from accessing decent housing and formal job opportunities, resulting in the deprivation problems of the residents in the obsolescent inner cities of urban China (Zhou et al. 2016). Furthermore, land finance and urban redevelopment policies in the cities of urban China are the root causes of the suburbanisation of poverty and the accessibility problems that are inherent in the jobs-housing spatial mismatch. For this reason, the accessibility constraints of low-income workers are always caused by social structures. These kinds of investigations should be investigated in terms of the concept of transport-related social exclusion.

2.5 The Concepts of Social Exclusion

Social exclusion is defined as a process of interaction between structure and agency that produces a state in which individuals are obstructed or unable to participate fully in the economic, social, political and cultural life in society (Giddens 1984). In addition, social exclusion is a social phenomenon in which individuals and groups in the population are deprived of the resources, freedom, goods and services necessary to achieve well-being. As a result, they are excluded from participating in mainstream activities in the societies to which they belong (Sen 2000; Levitas et al. 2007; Townsend 1979). Therefore, social exclusion tends to be understood as (United Nations 2016; Schwanen et al. 2015; Fuchs 2003) (Fig. 2.3):

- A lack of participation in social, economic and political activities that are important factors for an average standard of living;
- A self-organisation process in which the structuration theory is the mechanism of the process;

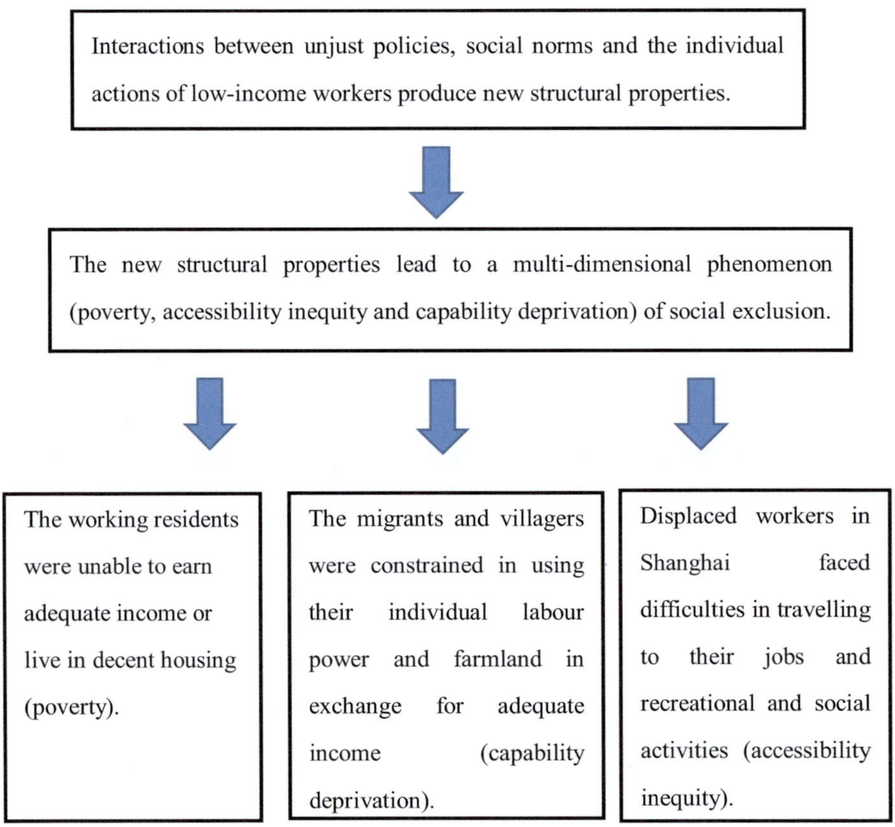

Fig. 2.3 Self-organisation process produces a multi-dimensional phenomenon of social exclusion

- Multidimensional and cumulative: those who are deprived of resources experience commuting problems in fulfilling basic needs and are excluded from participating in the mainstream activities of society because of a lack of human rights, freedom, choice and political power;
- Relative to other individuals or groups;
- Dynamic: it changes over time; and
- Multi-scalar: it affects individuals, neighbourhoods and local communities.

For example, high GDP growth is always the yardstick of local government performance in urban China. To generate adequate revenue to meet the targets set by the central government, many local governments, such as those of Shanghai and Guangzhou, carry out large-scale urban redevelopment to make land available for new commercial buildings. The local government can earn revenue from land sales and boost GDP growth (Tian et al. 2017a, b, c; Li and Wu 2010). The surge in market housing prices has made many low-income workers feel that they are unable to afford urban housing. They have to relocate to the urban periphery. Many of these displaced workers cannot find new jobs in the new towns and therefore choose to retire early. The impacts of high GDP growth, land finance and suburbanisation policies on displaced workers is the multi-dimensional phenomenon in this case of social exclusion.

2.5.1 Social Exclusion Includes Entitlement Failure

Preston and Rajé (2007) suggest that social exclusion is not due to a lack of employment, education and social welfare opportunities but to a lack of equitable government policies to ensure that the poor has access to those opportunities, which is due to unfair government policies and social norms. Sen (1980, 1981) finds that the famine that killed two to three million people in the 1943 Bengal Famine was triggered not by a decline in food availability but by speculation that arose from concerns about an impending Japanese invasion. Thus, as a result of the unjust trade policy, low-income workers could not exchange their labour power for adequate food.

Sen (1981) proposes the exchange entitlement theory to investigate famine and social exclusion. The theory stresses that unfair government policies exclude low-income workers from using their endowment set (e.g., labour-power, farmland) to make exchanges for adequate food and commodities (e.g., entitlement set) through exchange entitlement mapping (e-mapping), which is the exchange rate in open markets that workers can use with their endowment set in exchange for food and commodities. This failure of exchange entitlements is also called entitlement failure. Entitlement is a human right and is a form of social inclusion (Preston and Rajé 2007). The author adopts this theory for the social exclusion investigation, in which unjust policies (e.g., the *hukou* system and dualistic land-use system in China) and entitlement failure are interrelated. In other words, the unjust policies interact with the individual actions of low-income workers, resulting in entitlement failure. For example, the local governments in urban China acquire farmland for urban development, and the farmers in the suburbs cannot use their farmland to grow crops in exchange for income to maintain their quality of life. The farmers face entitlement failure, and the land acquisition policy excludes the farmers from using their labour power and farmland to achieve well-being. In addition, the *hukou* system constrains rural migrants in urban China from finding formal jobs and public housing, and rural migrants cannot exchange their labour power for food and income to maintain their quality of life. They face entitlement failure, and the *hukou* policy excludes rural migrants from participating in formal job markets in the host cities. Thus, social inclusion includes entitlement failure.

2.5.2 Transport-Related Social Exclusion

Regarding transport-related social exclusion, Church et al. (2000) suggest several factors that cause social exclusion:

- Physical exclusion is related to the physical and psychological barriers people face in accessing the transport system and the built environment;
- Geographical exclusion is related to the poor provision of transport and the resulting inaccessibility;
- Exclusion from facilities is related to a lack of access to good shopping, financial, recreational, health and education facilities;
- Economic exclusion is related to the capability to access employment opportunities;
- Time-based exclusion is related to the time poverty of working mothers; and
- Fear-based exclusion is related to modal and spatial actual and perceived safety.

Most transport studies on social exclusion are limited to the interactions between social groups, the built environment, the performance of transport systems and government policies (Xia et al. 2016; Ricciardi et al. 2015; Lucas 2012; Sen 2000). In 2003, the Social Exclusion Unit in the UK published a now internationally recognised report on social exclusion. The report resulted in the development of a set of transport policies in England (Lucas 2012; Social Exclusion Unit 2003). Policies on transport-related social exclusion mainly focus on the role of transport, land-use planning and service delivery decisions in improving the mobility of the working poor and elderly and handicapped persons. For example, a study (Xia et al. 2016) on transport-related social exclusion stated that:

> Transport-related social exclusion studies can be divided into two types based on the focus of research: category approach and spatial approach. The focus of the category approach is on the transport demand side, addressing issues such as travel patterns, attitudes and needs of particular social groups. The spatial approach emphasizes the transport supply-side, particularly the quality of transport supply, access to public transport, and spatial gaps of transport supply based on residents needs. (Xia et al. 2016, p. 205)

Focusing on transport systems and land-use development alone will make transport planners fail to understand the whole picture of the problems in transport-related social exclusion. Transport planners should also observe the influence of individual actions on travel behaviour. This is because structure (e.g., transport rules) is the medium and outcome of travel practices (Giddens 1981, 1984). Thus, interactions between structures and practices constitute social systems, including new behaviours.

For example, the travel problems of low-income working mothers involve complex interactions among various household responsibilities, employment, individual decisions about the locations of jobs or housing, and time allocations time employment and household duties (Blumenberg 2004). Therefore, transport studies on the social exclusion process should include individual decisions and actions, social norms and social networks, which influence travellers' routine travel behaviours (Giddens 1984; Sen 2000).

2.5.3 Government Policy Is the Root Cause of Transport-Related Social Exclusion

The transport literature indicates that social exclusion always stems from unfair government policies that exclude low-income workers from commuting to needed activities (Xia et al. 2016; Schwanen et al. 2015; Lucas 2012; Preston and Rajé 2007; Sen 2000). Social exclusion policies prevent low-income workers from having the capability to achieve well-being or from participating in the mainstream activities of society (Sen 2000). Such exclusion takes place when social structures, such as transport policies and the built-environment, influence low-income workers in such a way that the result is commuting problems and social exclusion.

For example, in Hong Kong SAR, the government-led public housing suburbanisation policy has relocated many low-income workers from urban areas to public housing estates in distant new towns. The highest number of job opportunities in Hong Kong SAR are located in the urban centre. As a result, many displaced low-income workers are excluded from accessing employment opportunities in urban areas, and they have difficulty finding new jobs in nearby new towns. Many of them become unemployed (Hui et al. 2015).

2.5.4 Social Exclusion and the Self-Organisation Process

Social exclusion is a self-organisation process that involves the impacts of socio-economic and land-use variables on low-income workers. The above structural variables interact with the variables of individual actions to produce new structural properties that result in social exclusion. The results of the studies in Chaps. 4 and 6 indicate that social exclusion cases involve multi-dimensional phenomena, such as the dimensions of poverty, accessibility inequity and capability deprivation (United Nations 2016; Sen 1984, 2000). These dimensions are as follows:

- Economic and housing reforms interacted with the decision of local residents to subdivide old houses in the historic inner city in Guangzhou into several rooms, resulting in the exclusion of the local residents from having a decent living environment (poverty dimension);
- The *hukou* system interacted with the decision of rural migrant workers to find affordable housing and informal jobs in the study urban village, resulting in the formation of a migrant enclave in Guangzhou. The *hukou* system caused rural migrants to face difficulty in using their labour power in exchange for an adequate income (capability deprivation and poverty dimensions);
- A concentration of migrant workers in the study urban village. Their social, employment and recreation activities were limited within the study urban village. The process resulted in the exclusion of migrants from reaching distant employment and non-work activities (accessibility inequity dimension);
- The gendered division of household labour in migrant households interacted with the decisions of working mothers to seek jobs near homes, resulting in the exclusion of the working mothers from reaching distant job opportunities (accessibility inequity dimension);
- Suburbanisation and land finance policies in Shanghai interacted with the decision of relocated workers to live in suburban areas, resulting in many displaced workers failing to find new jobs in the suburbs (capability deprivation and accessibility inequity dimensions); and
- The acquisition of farmland and the expansion of urban sprawl in Shanghai interacted with the decision of migrant workers to live close to their places of employment and take short-distance trips. The new travel behaviours caused the low demand for long-distance rail services (accessibility inequity dimension).

The multi-dimensional phenomenon of social exclusion in the case studies of this book is depicted in Fig. 2.3.

Furthermore, in the above cases, the self-organisation processes produce social exclusion and help transport planners identify accessibility problems. Neto et al. (2015) find that in Brazil, working mothers, both those who are single and those who are formerly married women, are prone to having shorter commutes than single fathers. This might be due to the interaction between social norms and the individual decisions of the women. In Brazil, men tend to be socially dominant, and women are expected to be self-sacrificing and submissive. Additionally, in her transport study on single mothers in the US, Blumenberg (2004) found that:

> Employment is not possible unless single mothers find appropriate childcare for their young children; are able to purchase groceries, clothes, and other necessities for the household; can attend the educational and health needs of their children and have the ability to respond quickly to family emergencies as they arise. (Blumenberg 2004, p. 277)

Thus, the shorter travel distances of women compared to those of men include many complex interactions between structures and actions to produce new behaviours, which can be investigated by means of a self-organisation approach.

Based on the results of in-depth interviews in Beijing, a study on the commuting of rural migrant workers (Li and Zhao 2018) found that:

> Their (rural migrants') work location was near their residence, and they seldom made recreation and social trips. One of the major reasons for this was that their social ties were highly limited to people from the same hometown. Additionally, time and financial poverty further limited their mobility. Institutional barriers also segregated migrant workers with local people in the job market, urban welfare and daily life. In order to improve physical mobility, migrant workers used the electric bicycle as their daily travel facility. (Li and Zhao 2018, p. 3)

According to the above examples, the accessibility problems of rural migrant workers are constrained by various structural factors and individual actions in social systems, and their complex travel behaviours can be investigated by the self-organisation approach.

2.6 Transforming Unjust Policies to Minimise Social Exclusion

The self-organisation process can be represented by recursive interactions between structure and agency. According to the duality of structure, the structural properties of social systems are both the medium and the outcome of the practices that constitute those systems (Giddens 1981). Structure and agency are a duality; neither one can exist without the other. Hence, structure and agency are two interdependent variables. Individual decisions and actions reproduce and transform social norms, values and government policies through routine practices. For example, working mothers have to take on more household responsibilities than do working fathers because of the norms that allocate household labour in two-worker households; working mothers face time conflicts between their household responsibilities and employment. Social norms exclude the working mothers from participating in the mainstream activities of society. At the same time, the individual decisions and actions of some working mothers transform the social norms through the reproduction of their practices in achieving employment. Some working mothers find part-time jobs and jobs near their homes so that they can cut working hours and travel times and make more time for childcare and recreational activities. If working mothers find new travel practices to tackle time conflict problems and improve their access to jobs, government decision-makers can also change policies to improve accessibility and minimise social exclusion.

The above case demonstrates that government decision-makers can tackle the problems of transport-related social exclusion by transforming unjust government policies to improve accessibility for low-income workers. For example, single mothers may feel that employment is not possible unless they find appropriate childcare for their young children and are able to purchase groceries, clothes, and other necessities for their household. Government decision-makers could provide childcare subsidies for single mothers. In addition, policy-makers could launch a new policy of admitting foreign domestic helpers and subsidise single mothers to employ domestic helpers to reduce their household duties.

2.7 Enabling Individual Actions to Minimise Social Exclusion

Social structures are capable of both constraining and enabling action (Giddens 1984). Structures often allow actors to do things that they would otherwise not be able to do. Public participation is a strategy for policy-makers to understand individual actions and attitudes so that they can modify policies to enable individual actions and minimise social exclusion. Lucas (2012) suggests that policy-makers can minimise social exclusion if they recognise the abilities, skills, resources, capacities and experiences of affected individuals in the design of policy solutions. Policy-makers can formulate appropriate policies to help affected individuals so that the latter have more freedom and choice to achieve well-being. Hence, policy-makers have to listen to the travel needs of affected individuals, and there is a need to look at the transport planning process to ensure that the consequences of transport-related social exclusion are better communicated and understood.

The collaborative approach is introduced as a strategy to meet the demands of the communist ideology and improve social inclusion and personal commuting in communist societies such as China (Healey 1997). Thus, policy-makers should respect all stakeholders in the process of policy formulation, and government departments should promote public debate on government policies so that policy-makers can listen to the affected individuals about their actions and attitudes (e.g., they spend long hours working in informal labour markets and generate a low trip frequency per day or low trip frequency to non-work activities). Policymakers should modify the unjust policies to enable the action of affected workers and to minimise social exclusion.

Harvey (1996) argues that during the process of urban development, policy-makers should make room for diversity, difference, and the heterogeneity of values and lifestyle oppositions. Cities that cannot accommodate diversity, migratory movements, new lifestyles and new economic, political, religious values and embrace heterogeneity will die either through stagnation or because they will fall apart in violent conflict.

Healey (1997, 1998) proposes the collaborative approach as a strategy for encouraging public participation. The collaborative approach includes several steps. First, government cadres should respect the rights of all stakeholders to voice their views on government policies. Second, cadres should treat low-income workers and rural migrants as stakeholders. Third, the Chinese government should impose a duty on all senior cadres to attend to the concerns of all stakeholders. Finally, the collaborative approach includes two types of institutional infrastructures, namely, hard (e.g., consultation through district governments and street offices) and soft (e.g., consultation through radio and television programmes and readers' columns in local newspapers) infrastructure. This method requires government cadres to break out of traditional hierarchical and 'bureaucratic' processes to involve new groups, networks, and 'partnerships', including with affected low-income workers.

Certainly, government decision-makers who want to build a just society must tolerate the complexity of the encounters between people with different values, frames of reference and ways of organising practices.

2.8 Conclusion

Chapter 2 reviews the concept of the self-organisation process, travel behaviours produced by the self-organisation process and the concept of social exclusion. Self-organisation is a process in which a human social system reproduces itself with the help of interactions between the structures and actions of individual human actors to maintain balanced development and constitute the social system. The self-organisation process produces new structural properties from the old social system, and these new structures include new travel behaviours that lead to accessibility problems and social exclusion. The structuration theory is the mechanism of the self-organisation process, and the main concept of the theory is based on the duality of structure, which means that structures are both the medium and the outcome of the practices (Giddens 1981, 1984).

Self-organisation in a broad sense refers to the re-creation of society. In this book, self-organisation focuses on government policies and social norms that constrain or enable low-income workers in accessing resources for basic functioning, such as housing and employment. Structures exist in the form of top-down emergence, such as the *hukou* system and social networks, and the actions of low-income workers exist in the form of bottom-up emergence. The actions of low-income workers transform the structures through the reproduction of routine social interactions. In such a way, at least one new structural property is produced and recreates the social system. This book focuses on new structural properties that are related to the creation of new behaviours that are themselves related to employment and non-work activities.

Transport planners always fail to see the whole picture of travel behaviour in their efforts to minimise social exclusion. Transport planners should consider the influence of individual decisions and actions on travel behaviours. This is because structure (e.g., transport system) is both the medium and the outcome of travel practices (Giddens 1981, 1984). Thus, the interaction between structure and practice constitutes social systems, which include new travel behaviours. Transport planners can understand individual actions by paying attention to the needs, skills and capability of affected low-income workers. For example, the travel problems of low-income working mothers involve complex interactions between individual decisions and structures. The gendered division of household labour allocates more household responsibilities to women than to men. This social norm interacts with the individual decisions of working mothers in choosing working locations near their homes and produces short-distance commutes. Therefore, the self-organisation process, which is also the cycle of agency and structure, is identified as an effective research tool to investigate travel behaviour.

In addition, this chapter investigates the concept of social exclusion and the relationship between transport-related social exclusion and the self-organisation process. Social exclusion is a multi-dimensional phenomenon that prevents low-income workers from achieving well-being due to resource deprivation, commuting problems and a lack of freedom, choice and opportunity to participate in the mainstream activities of society (United Nations 2016; Sen 1984, 2000; Rawls 1971). Social exclusion is also a form of entitlement failure in which unjust policies in urban China exclude low-income workers from participating in formal labour markets and finding decent housing in coastal cities so that the low-income workers cannot exchange their labour power for adequate food and income.

Social exclusion is also a self-organisation process, and the mechanism of the process is the duality of structure. Structures, such as social norms and unjust government policies, interact with individual decisions and actions to produce new travel behaviours that result in inaccessibility and social exclusion. Structure and agency are two interdependent variables, and policy-makers can minimise the impacts of social exclusion by changing either the structure or individual actions. This chapter proposes that decision-makers who want to build a just society in urban China can use social justice theories to deliver equitable policies for minimising social exclusion and improving accessibility. For example, government decision-makers should consider entitlement theory, which stresses that government policies should ensure that low-income workers can exchange their labour power and farmland for adequate food and wages. Thus, the *hukou* system should be amended in accordance with entitlement theory.

In addition, policy-makers in the coastal cities of China can minimise social exclusion through public participation in the process of policy formulation. If policy-makers can understand the needs and difficulties of the affected low-income workers, they can formulate appropriate policies to enable individual actions and minimise social exclusion, such as providing appropriate policies for poor residents so that they can have more freedom and choice to achieve well-being.

References

Ball P (2013) Why society is a complex matter. Springer, Berlin

Batty M (2009) Cities as complex systems: scaling, interactions, networks, dynamics and urban morphologies. In: Meyers R (ed) Encyclopedia of complexity and urban systems science. Springer, Berlin De

Beyazit E (2011) Evaluating social justice in transport: lessons to be learned from the capability approach. Transp Rev 31(1):117–134

Blumenberg E (2004) En-gendering effective planning: spatial mismatch, low-income women, and transportation policy. J Am Plan Assoc 70(3):269–281

Church A, Frost M, Sullivan K (2000) Transport and social exclusion in London. Transp Policy 7 (3):195–205

De Roo G (2016) Self-organization and spatial planning: foundations, challenges, constraints and consequences. In: De Roo G, Boelens L (eds) Spatial planning in a complex unpredictable world of change. InPlanning, Groningen

Dimitriou HT (2013) Transport planning for third world cities (Routledge revivals). Routledge, London

Fuchs C (2002) Concepts of social self-organisation, INTAS Project "Human Strategies in Complexity" Research Paper No. 4. Available at SSRN: https://ssrn.com/abstract=385185. Accessed 10 Oct 2018

Fuchs C (2003) Structuration theory and self-organization. Syst Pract Action Res 16(2):133–167

Fuchs C, Hofkirchner W (2005) The dialectic of bottom-up and top-down emergence in social systems. TripleC i(i):28–50

Giddens A (1981) A contemporary critique of historical materialism. Vol. 1. Power, property and the state. Macmillan, London

Giddens A (1984) The constitution of society. Polity Press, Cambridge

Harvey D (1996) Justice, nature and the politics of difference. Blackwell, Oxford

Healey P (1997) Collaborative planning: shaping places in fragmented societies. UBC Press, Vancouver

Healey P (1998) Collaborative planning in a stakeholder society. Town Plan Rev 69(1):1–21

Heylighen F (2001) The science of self-organisation and Adaptivity. The encyclopedia of life support systems. Free University of Brussels, Brussels

Heylighen F (2008) Complexity and self-organization. In: Bates MJ, Maack MN (eds) Encyclopedia of library and information sciences, 3rd edn. Taylor & Francis, London, pp 1215–1224

Hui ECM, Zhong JW, Yu KH (2015) Housing policy, work-residence mismatch and poverty concentration. Habitat Int 48:198–208

Jiading District Statistics Bureau (2014) Analysis on the survey results about living conditions of rural migrant workers in Jiading District, Jiading District, Shanghai Municipal Government. http://www.stats-sh.gov.cn/html/fxbg/201402/267398.html. Accessed 7 June 2018 (in Chinese)

Kloch K, Pirkl G, Lukowicz P, Fischer C (2011) Emergent behaviour in collaborative indoor localisation: an example of self-organisation in ubiquitous sensing systems. In: Berekovic M, Fornaciari W, Brinkschulte U, Silvano C (eds) Architecture of computing systems – ARCS 2011. ARCS 2011. Lecture notes in computer science, vol 6566. Springer, Berlin

Levitas R, Pantazis C, Fahmy E, Gordon D, Lloyd E, Patsios D (2007) The multi- dimensional analysis of social exclusion, department of sociology and school for social policy, Townsend centre for the international study of poverty and Bristol institute for public affairs. University of Bristol, Bristol

Li ZG, Wu FL (2010) Post-reform residential segregation in three Chinese cities: Beijing, Shanghai and Guangzhou. In: Wu FL, Webster C (eds) Marginalization in urban China. Palgrave Macmillan, London, pp 226–250

Li SX, Zhao PG (2018) Restrained mobility in a high-accessible and migrant-rich area in downtown Beijing. Eur Transp Res Rev 10(4):2–18

Lucas K (2012) Transport and social exclusion: where are we now? Transp Policy 20:105–113

McQuoid J, Dijst M (2012) Bringing emotions to time geography: the case of mobilities of poverty. J Transp Geogr 23:26–34

Mingers J (1995) Self-producing systems. Implications and applications of Autopoiesis. Plenum Press, New York

Morrison K (2005) Structuration theory, habitus and complexity theory: elective affinities or old wine in new bottles? Br J Sociol Educ 26(3):311–326

Mote JE, Whitestone Y (2011) The social context of informal commuting: slugs, strangers and structuration. Transp Res A 45:258–268
Neto RS, Duarte G, Páez A (2015) Gender and commuting time in São Paulo metropolitan region. Urban Stud 52(2):298–313
Ouyang W, Wang B, Tian L, Niu X (2017) Spatial deprivation of urban public services in migrant enclaves under the context of a rapidly urbanizing China: an evaluation based on suburban Shanghai. Cities 60:436–445
Portugali J (2000) Self-organization and the city. Springer, Heidelberg
Portugali J (2006) Complexity theory as a link between space and place. Environ Plan A 38:647–664
Portugali J (2017) Self-organization and the city. In: Meyers R (ed) Encyclopaedia of complexity and systems science. Springer, Berlin
Preston J, Rajé F (2007) Accessibility, mobility and transport-related social exclusion. J Transp Geogr 15:151–160
Rauws W, De Roo G, Zhang S (2016) Self-organisation and spatial planning: an editorial introduction. Town Plan Rev 87(3):241–251
Rawls J (1971) A theory of justice. Oxford University Press, Oxford
Ricciardi AM, Xia JC, Currie G (2015) Exploring public transport equity between separate disadvantaged cohorts: a case study in Perth, Australia. J Transp Geogr 43:111–122
Schwanen T, Lucas K, Akyelken N, Solsona DC, Carrasco JA, Neutens T (2015) Rethinking the links between social exclusion and transport disadvantage through the lens of social capital. Transp Res A 74:123–135
Sen A (1980) Equality of what? Choice, welfare and measurement (Cambridge, MA: Harvard University Press), 1997 [Originally published in The Tanner Lectures on Human Values, Vol. I, by the University of Utah Press and Cambridge University Press, 1980]
Sen A (1981) Poverty and famines: an essay on entitlement and deprivation. Clarendon Press, Oxford
Sen A (1984) Resources, values and development. Oxford, Basil Blackwell
Sen A (1992) Inequality reexamined. Oxford University Press, Oxford
Sen A (1999) Development as freedom. Oxford University Press, Oxford
Sen A (2000) Social exclusion: concept, application and scrutiny. Social development papers no 1. Office of Environment and Social Development, Asian Development Bank, Manila
Smith A (1937) An inquiry into the nature and causes of the wealth of nations [1776]. In: Cannan E (ed) . Random House, New York
Social Exclusion Unit (2003) Making the connections: final report on transport and social exclusion. Office of the Deputy Prime Minister, United Kingdom, London
Tian L, Ge B, Li Y (2017a) Impacts of state-led and bottom-up urbanization on land use change in the peri-urban areas of Shanghai: planned growth or uncontrolled sprawl? Cities 60:476–486
Tian L, Guo X, Yin W (2017b) From urban sprawl to land consolidation in suburban Shanghai under the backdrop of increasing versus decreasing balance policy: a perspective of property rights transfer. Urban Stud 54(4):878–896
Tian L, Li Y, Shao L, Zhang Y (2017c) Measuring spatial-temporal characteristics of city expansion and its driving forces in Shanghai from 1990 to 2015. Chin Geogr Sci 27(6):875–890
Townsend P (1979) Poverty in the United Kingdom: a survey of household resources and standard of living. University of California, pp 31–60
United Nations (2016) Identifying social inclusion and exclusion. In: Leaving no one behind: the imperative of inclusive development, report on the social situation 2016. Department of Economic and Social Affairs, New York, pp 17–28
Xia JC, Nesbitt J, Daley R, Najnin A, Litman T, Tiwari SP (2016) A multi-dimensional view of transport-related social exclusion: a comparative study of greater Perth and Sydney. Transp Res A 94:205–221
Zhang SH, De Roo G (2016) Interdependency of self-organisation and planning: evidence from Nanluoguxiang Beijing. Town Plan Rev 87(3):253–274

Zhou SH, Liu Y, Kwan MP (2016) Spatial mismatch in post-reform urban China: a case study of a relocated state-owned enterprise in Guangzhou. Habitat Int 58:1–11

Zhu PY, Zhao SN, Wang LP, Yammahi SA (2017) Residential segregation and commuting patterns of migrant works in China. Transp Res D 52:586–599

Chapter 3
Self-Organisation Process, Accessibility and Principles of Justice

3.1 Introduction

This chapter focuses on investigating the roles of accessibility and social justice theories in the self-organisation process. Accessibility is a concept in transport planning that is used to measure the ability of an individual to penetrate constraints to reach his desired activities. In the studies of the self-organisation process in the following chapters, the accessibility problems of low-income workers in urban China in overcoming constraints on reaching employment opportunities and non-work activities will be introduced. The low-income workers in the deprived neighbourhoods being studied experience self-organisation processes in which they use their individual decisions and actions to interact with the various government policies and social norms to achieve employment and housing. Many of them have to work long hours or face time conflicts between household duties and employment. As a result, they choose to spend less time pursuing non-work activities, such as social and recreational activities.

Hence, the concept of accessibility helps measure the impacts of the new travel behaviours, which are produced by the self-organisation process, on quality of life. For example, in an urban village where the self-organisation process takes place, the concepts of space-time prisms and potential path areas are used to measure the accessibility of poor workers under the effects of space-time constraints. In addition, accessibility concepts can help transport planners understand how the self-organisation process produces new travel behaviours that result in social exclusion problems.

In the first part of this chapter, the concepts of accessibility that are based on travel needs and the capability approach will be introduced. This is followed by an explanation of the self-organisation process that includes individuals' decisions to trade non-work needed activities for employment and trade travel time for longer working hours. A space-time prism approach is used to explain the above trade-off

J. C.-y. Lau, *Self-Organisation Shapes Travel Behaviours and Social Exclusion in Deprived Urban Neighbourhoods of China*,
https://doi.org/10.1007/978-981-15-2252-9_3

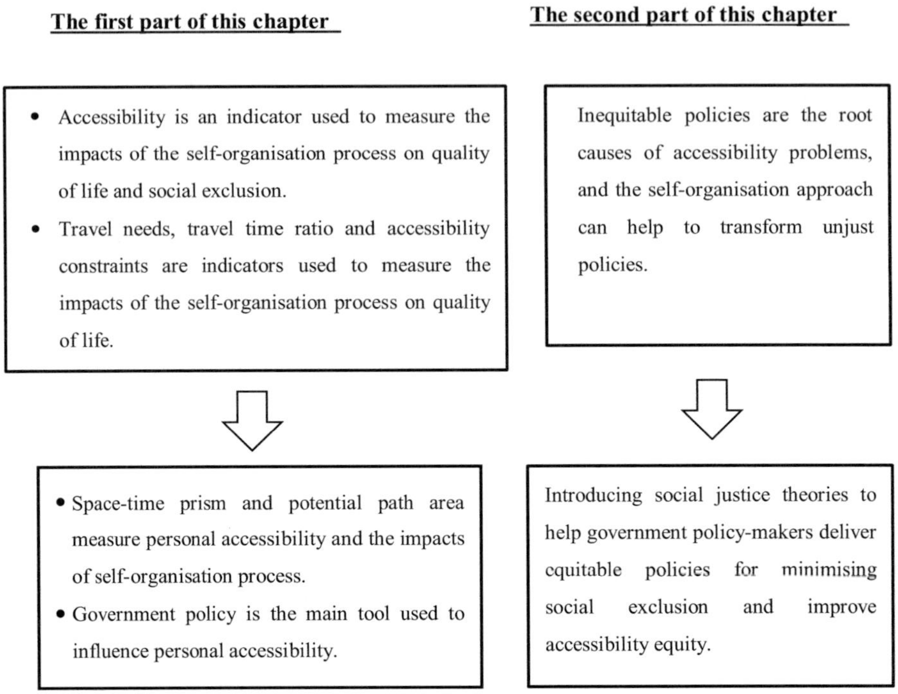

Fig. 3.1 The roles of accessibility and justice theories in the self-organisation process

decisions. This indicates that the trade-offs involved in individual decisions helps explain why a self-organisation process produces accessibility problems for low-income workers (Fig. 3.1).

The second part of this chapter introduces major social justice theories. The structure is both the medium and the outcome of the reproduction of practices (Giddens 1981, 1984). Thus, policy and individual interaction constitute social systems that include new structural properties. It is assumed that unfair policies are the root cause of accessibility problems. Unjust policies interact with agency in a self-organisation process, thus producing new structural properties that lead to social exclusion. Structure and action are independent variables, and thus decision-makers who want to build a just society can transform either policies or the individual actions in the self-organisation process to minimise social exclusion. Chapter 3 investigates the roles of social justice theories in helping decision-makers who want to build a just society modify unjust policies to improve the quality of life and accessibility of low-income workers (Fig. 3.1).

3.2 Accessibility and Its Role in the Self-Organisation Process

The major role of accessibility in the self-organisation process is to measure and identify the impacts of the self-organisation process on the commuting of low-income workers to reach needed activities. In the transport literature, there are several ways to measure accessibility, and this section investigates the accessibility definitions and measures that are suitable for studying the self-organisation process.

The transport literature always measures land-use and transportation performance by means of accessibility. Accessibility is an important tool for cost and benefits analysis in assessing transportation networks, the land-use system in a given region and access to opportunities (Boisjoly and El-Geneidy 2017). Five of the most commonly used definitions in the literature are (Boisjoly and El-Geneidy 2017; Geurs and Wee 2004):

- The distance (or travel time) to the nearest desirable location, such as freeway interchange and employment opportunities;
- The cumulative opportunities which count all opportunities that can be reached within a travel cost threshold (isochrone method);
- The potential of opportunities for interaction (the gravity-based measures) (Hansen 1959; Meyer and Miller 2001);
- The overall benefits provided by a given transport system (Ben-Akiva and Lerman 1979); and
- The activity space of an individual within a particular space-time context (prism) (Zhang et al. 2019).

The above five accessibility definitions are mostly based on the impacts of policies and social factors on commuting. For example, the policies of building transport systems to improve public transit service frequency, reducing transport costs, providing a wider array of choices for transport modes (for example, walking, public transport and private cars) and increasing network density were implemented to enhance the potential number of opportunities for interaction. However, the above definitions of accessibility ignore two important factors. First, the influence of individual decisions and actions on accessibility.

According to Giddens, every human agent is knowledgeable in the practical consciousness and has a vast variety of tacit modes of knowing how to go on in the contexts of social life (Giddens 1979). Hence, agents use their knowledge to diagnose the environment and choose their travel actions. Miller (2018) suggests that:

> The ability of people to exploit a given physical context of access and attraction will also vary depending on their financial, cognitive and physical capabilities. As obvious examples, persons without cars and persons with physical and mental disabilities all generally have quite different accessibilities (potential for interactions) than persons who do not face these same challenges. Therefore, accessibility ultimately varies among persons as a function of both their individual preferences and capabilities. (Miller 2018, p. 552)

In addition, a study in Munich found that symbolic Advanced- Producer-Services (APS) workers tend to reside in central areas and use public transport to commute. In contrast, synthetic high-tech workers are found in relatively peripheral areas and depend more on cars to reach their workplaces. The spatially-related choices of analytical high-tech and synthetic-APS workers are positioned in between symbolic APS-workers and synthetic high-tech workers (Zhao et al. 2017). The results of the study indicated that the features of the knowledge base are evident in the spatial choices of knowledge workers. In addition, while the influence of the knowledge base has to be weighed against socio-demographic factors, different groups of knowledge actors clearly display distinct choices of residential location and commute mode.

Second, the most commonly used definitions in the literature have no objective, normative standard for what constitutes 'good' or 'acceptable' accessibility (Miller 2018). Using an isochrone measure, for example, what is the value of access to one more job within a 30-min contour timeline to an individual's quality of life? To fill this research gap, this book introduces a good accessibility definition that defines accessibility as freedom to achieve well-being.

3.3 Accessibility Means Freedom to Achieve Well-being and the Ability to Penetrate Obstructions

Accessibility is defined as the ability of an individual to overcome constraints to achieve well-being and reach the mainstream activities of his or her society. The definition involves two important concepts, the term 'ability' is linked to the right to move freely to access activities and the term 'the mainstream activities of society' is linked to capability and social exclusion (Van Wee and Geurs 2011).

The right of an individual to move about freely to reach needed activities can be explained by using the capability approach (Sen 1984; Cao and Hickman 2019; Ryan et al. 2019). According to the capability approach, the operationalised capabilities are referred to as the individual's functionings (e.g., the individual's realized travel behaviour to employment) (Ryan et al. 2019). Sen suggests that a government characterised by justice and equity should provide all people with the freedom, choice and opportunity to achieve well-being. In other words, people should have adequate time to travel to and stay at needed activities. People should have adequate freedom, choice and opportunities to choosing how to reach their needed activities, including using different transport modes, travel distances and trip rates to reach jobs and non-work activities. For example, rural migrant workers in urban China should have freedom and choice to choose their recreational and employment activities; they should have adequate choices and opportunities to choose their housing and social activities. The Chinese government should modify the *hukou* system to improve the accessibility of rural migrants in these aspects.

The concept of accessibility is also related to the difference principle suggested by Rawls (1971). Rawls states that accessibility inequities are features of human societies, but the inequities of accessibility can actually be just, as long as they benefit the least well off in society, such as rural migrants in urban China. Therefore, an equitable transport policy should ensure that all people, particularly low-income workers, have freedom and choice in reaching different jobs and non-work opportunities. In this study, the focus will be on government policies that hinder low-income workers in commuting to their places of employment and non-work activities; one such policy is the *hukou* system, which constrains migrant workers from commuting to formal labour markets and social and recreational activities.

In addition, one of the criteria of the accessibility definition is related to individuals who experience social exclusion as a result of the obstructions imposed on them that hinder them in commuting to the mainstream activities of their society. The built environment and transport systems are the main barriers that cause obstructions in commuting to places to fulfil basic needs (Sen 2000; Levitas et al. 2007). In urban China, the spatial development and suburbanisation of cities always causes jobs-housing spatial mismatches and an inadequate provision of transport services for workers in suburban areas. Accessibility problems include workers who are constrained by transport systems, unaffordable public transport fares, travelling long distances and living far away from employment opportunities.

This book will use the concept of accessibility to measure the new travel behaviour produced by the self-organisation process, as seen in Chap. 6. This book uses the accessibility concept to explain the jobs-housing spatial mismatch problems faced by displaced low-income residents in the process of the suburbanisation of poverty in Shanghai.

Furthermore, in urban China, the policies of the suburbanisation and transit-oriented development of cities led to an increase in land values in the urban centre. Low-income workers, who cannot afford the high land prices, often decide to relocate to suburban areas. This interaction between land finance policies and individual actions produces new travel behaviours and accessibility problems, such as the building of a rail transit service to link the new towns and city centre and relieve traffic congestion on the main roads. In addition, transit services cannot cover all dispersed housing locations in the suburbs.

Furthermore, in China, the *danwei* system interacts with the actions of the workers and produces new travel behaviour, which can be evaluated by means of the concept of accessibility. The *danwei* system was a policy instituted in China before the 1980s in which all workers were grouped into basic units of economic, social and spatial organisation and economic activity, social life, and political control were integrated and located within a neighbourhood. These neighbourhoods were usually located in or near the inner core of cities. The *danwei* workers were meant to develop political, productive, recreational and social activities all within the walls of the *danwei* compounds. As a result, the interaction between the *danwei* system and the actions of workers produced new travel behaviours, which positively influenced their access to employment; for example, they generated short-distance daily trips and a large share of workers used non-motorised transport modes (Zacharias 2012; Wang and Chai 2009).

Accessibility can also be used to evaluate the self-organisation process that takes place in the development of inner cities. Many low-income workers decide to trade proximity to employment and the easy availability of transportation for an overcrowded living environment. They decide to live in deprived inner cities. For example, Chap. 5 describes a case study that was conducted to investigate the social exclusion and commuting problems of low-income residents in Xiguan, the deprived but historic inner city in Guangzhou. The results of the study indicated that the interaction between the housing allocation system and actions of the local residents produced new travel behaviours that positively influenced the accessibility of jobs. The residents in Xiguan took commuted short distances to reach work, and many of them used non-mechanical transport modes to reach their places of employment. Some of them could find jobs in other districts in Guangzhou, but they experienced accessibility problems because of the outdated road systems that caused conflicts between pedestrians and vehicle traffic.

3.4 Travel Needs and their Relationship to the Self-Organisation Process

Similar to accessibility, the concept of travel needs can be used as a tool to evaluate the new travel behaviours that are produced by the self-organisation process. Travel needs are the needs of individuals for transport tools or public transit services to commute to activities to satisfy their basic needs (e.g., employment, school, medical services and social participation). Individuals have different needs for transport services, and their travel needs always depend on their socio-economic backgrounds. Smith et al. (2012) provided a list of travel needs that low-income households have in driving private cars or taking public transit to participate in society to maintain their standard of living in the UK. The activities of members of these households included working, attending school, using playgrounds, shopping for food and household goods, participating in social and cultural activities, attending to personal business, and visiting doctors. The study found that travel needs varied for different household types in the UK. Low-income working adults, youths, elderly persons, and working mothers have different travel needs.

Unmet travel needs refer to mobility deprivation/the inability of a person to travel to a facility because of inadequate transit services. A study that investigated the commuting problems of low-income regions in the US suggested that travel needs are defined as a quantification of people in an area who are in need of public transportation services (e.g., the number, percentage). Thus, unmet travel needs denote the gap between the travel needs of a population group and the transport supply (Pyrialakou et al. 2016).

Other than transport supply, the fulfilment of a person's travel needs also depends on the level of his or her physical ability, which allows the person to walk or drive freely to reach employment opportunities, medical facilities and other social services. Furthermore, low-income workers may experience unmet mobility needs because they feel unable to afford public transit fares or car fuel prices to reach the needed activities.

The concept of travel needs can help transport planners evaluate the impacts of the self-organisation process on the commuting of affected low-income workers to places where they can satisfy their basic needs. For example, Chap. 4 investigates the self-organisation process that took place in the study urban village in Guangzhou. The results of the studies indicate that the *hukou* system interacted with the actions of the rural migrant workers in the study urban village to produce new travel behaviours, which caused commuting problems for the migrants in reaching social and recreational activities. Some of the migrant workers wanted to take long-distance trips to reach distant social activities in the city centre. However, the author observed that there were inadequate public bus routes to link the urban village with other districts in Guangzhou. Thus, the self-organisation process that took place in the study urban village produced new structural properties, which resulted in the travel needs of the rural migrants being unmet due to the inadequate provision of public bus services.

3.4.1 Travel Time Ratio and Individual Actions in Self-Organisation

The self-organisation process is operationalised by means of individual decisions, and actions that interact with structures. In the process, low-income workers' individual decisions modify structures through the reproduction of routine practices. In the transport literature, the travel time ratio is found to be a strategy through which individual decisions and actions modify structures to produce new travel behaviours. The travel time ratio is a concept of trading travel time for activity time in order to achieve one's intended goals, such as obtaining higher income. Thus, understanding an individual's decision on the trade-off between travel time and activity time when given a limited time budget is an important factor in understanding the self-organisation process and commuting behaviours.

Dijst and Vidakovic (2000) suggest the travel-time ratio (TTR), which is defined as the ratio of total amount of travel time to amount of time spent at an activity. Chen and Mokhtarian (2006) proposed the concept of travel-time price (TTP), which is defined as the ratio of travel time to activity time. Both of these ratios are used to investigate individual decisions regarding the trade-off between travel and activity time within a 24-h period of a given space-time prism (Hägerstrand 1970).

In Beijing, a study was conducted on the trade-off between travel and activity time for both mandatory and discretionary activities for multi-purpose commuting trips (Mao et al. 2018). Based on the concepts of TTR and TTP, the study found that a longer commute duration encourages shorter activity time for eating, shopping and family activities. In addition, a longer work duration reduces the time budget for travellers and consequently decreases their travel time for non-work stops, regardless of activity type. The findings of the above study help to explain the decisions of rural migrants to allocate time to commuting and employment activities in the self-organisation process that is taking place in the coastal cities of China. The process produces new travel behaviours, which include migrants' short–distance commutes in urban villages. One of the causes of the migrants' short-distance trip patterns is the long duration of their work obligations, which reduces their time budget for travelling for employment and their trip frequency to non-work activities. In addition, many medical, recreational and shopping facilities are located far away from migrant enclaves. Hence, migrants experience accessibility problems.

3.5 Accessibility Constraints Produced by the Self-Organisation Process

The mobility-based approach dominates transportation planning to mitigate accessibility constraints and improve accessibility (Boisjoly and El-Geneidy 2017). Many transport planners still maintain their minds that accessibility is the potential for interaction opportunities (Hansen 1959). A majority of the transport literature focuses on minimising accessibility constraints through improvements in the built environment, proximity among locations and the performance of transport systems, such as providing more road networks and public transport services, building mixed land-use development, and avoiding urban sprawl (Yang et al. 2017; Wang and Zhou 2017). For example, the development of high-speed train networks to reduce travel times between cities is a recent trend in transportation planning in China (China Daily 2017).

The mobility-based approach fails to measure constraints, which are related to social aspects. Improvements in transport systems cannot easily overcome accessibility constraints in deprived neighbourhoods, particularly affordability constraints. For example, poor residents experience accessibility constraints in satisfying basic needs because they feel unable to afford public transit fares. As a result, transport research points towards the inclusion of individuals' space-time constraints. Hägerstrand (1970) proposes three types of space-time constraints that influence an individual's accessibility to desired potential opportunities: coupling, authority and capability constraints (Neutens et al. 2007):

- Coupling constraints refer to where, when and how long an individual has to join other individuals to reach a destination;
- Authority constraints are institutionally imposed restrictions, such as those exerted by office hours and the service hours of public transport;

- Capability constraints are physical limitations on an individual's ability to access activities (such as the physical disability of an elderly person) and transportation modes (e.g., bicycle or walking).

The measurements of space-time constraints fail to reflect the situations in which individual decisions interact with policies or norms to produce accessibility constraints. For example, rural migrant workers face authority constraints that institutionally impose restrictions on them in accessing formal labour markets in urban China, e.g., the *hukou* system. The workers develop individual decisions to modify this accessibility constraint by gathering in urban villages in the inner suburbs of coastal cities, making a living as informal manufacturing workers and seeking informal jobs by means of their social network. Since the relevant accessibility constraints involve the *hukou* system, individual decisions and social networks, it is difficult to measure the constraints by conventional urban transport planning models.

3.5.1 Space-Time Prism Measures of the New Travel Behaviours Produced by Self-Organisation

Even though conventional accessibility measures (e.g., the space-time prism approach) cannot measure the whole self-organisation process that produces new travel behaviour, they can measure new travel behaviours (e.g., travel distance and travel time) and evaluate the impacts of the self-organisation process on the quality of life of the affected individuals. For example, the results of the studies in Chaps. 4 and 5 indicated that the self-organisation processes that took place in the deprived urban neighbourhoods produced new structural properties, which include the migrant workers taking short-distance trips to employment and non-work activities. The concepts of the space-time prism and potential path area can measure the accessibility levels of rural migrant workers who take short-distance trips. The accessibility measures also reflect that migrant workers suffer from social exclusion.

This section introduces the concepts of the space-time prism approach and potential path area and examines how travel behaviours, e.g., travel distance and travel time, are important variables in the evaluation of an individual's accessibility level. A space-time prism is the set of all activities that an individual can reach given space-time constraints, such as travel distance, travel speed and duration of the activities, within a 24-h period of a given space-time prism (Hägerstrand 1970) (Fig. 3.2).

The projection of the three-dimensional space-time prism onto a two-dimensional geographical space is called the potential path area (PPA), which is the spatial extent to which individuals can participate in their desired activities when subject to travel time and travel cost constraints (Patterson and Farber 2015) (Fig. 3.2). The concept of PPA is a measure of the outcomes of activity participation and their associated travel constraints with that activity. The PPA approach is a way to incorporate the

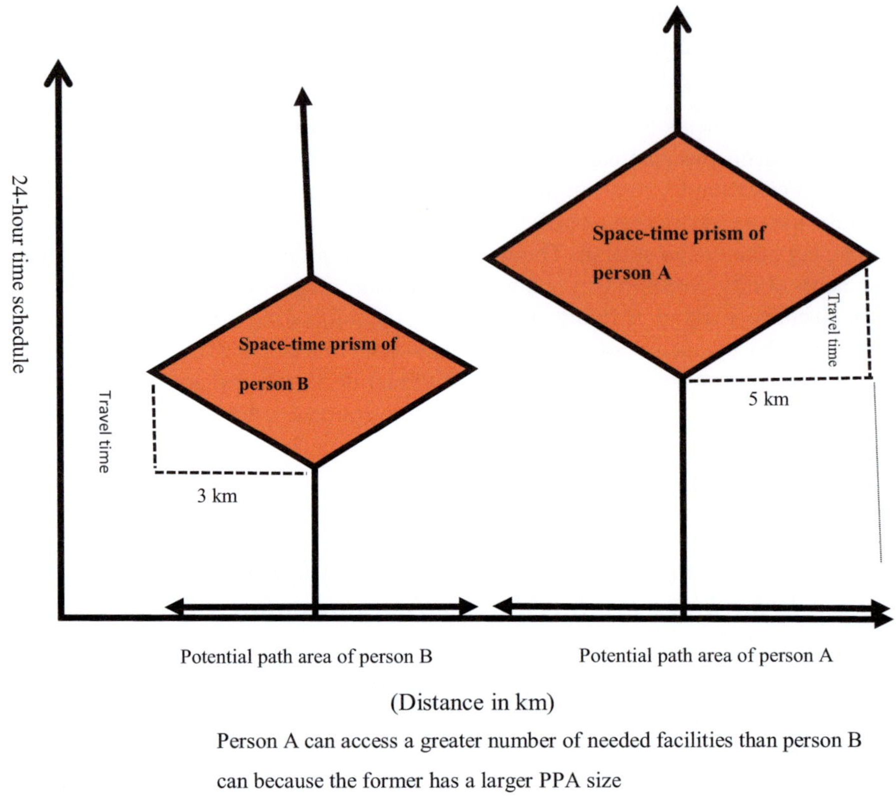

Fig. 3.2 To compare the different physical sizes and accessibility levels of PPAs between person A and B

issues of space, time and social accessibility constraints so that an individual can overcome the constraints of travel for activity participation.

An individual's accessibility is influenced by the distance he or she commutes to reach necessary employment and distant non-work opportunities. The longer the distance a person commutes, the larger the size of the PPA he or she has, and the higher the level of accessibility he or she experiences. The physical size of an individual's PPA is determined by travel distances, the amount of time available for commuting, the stay time at activities, and the speed of the transport modes (Miller 1991).

Kim and Kwan (2003) suggested that the physical size of a PPA is positively related to the average travel distance of a person to activities. In Fig. 3.2, the travel distance of person A is 5 km, while person B's travel time is 3 km. As a result, the physical size of person A's PPA is larger than that of person B, which implies that person A can reach a wider range of potential needed activities than person B can. Thus, the accessibility level of person A is higher than that of person B (Fig. 3.2). In such a manner, in the case studies in Chaps. 4 and 5 of this book, the rural migrant workers, who commute short distances, also experience accessibility problems in reaching distant non-work activities, such as social, medical and recreational activities.

In addition, a study examined the relationship between the characteristics of individual activity space and social segregation of the residents of different types of housing in Beijing (Zhang et al. 2019). The results of the study indicated that the residents of affordable housing have both the smallest actual and potential activity space on weekdays, which makes them the most isolated group when compared with other groups (e.g., the residents of market housing and institutional housing). This may indicate that the lower social status of the residents of affordable housing in relation to their actual daily use of space and their lower capacity of 'potential mobilities' as a form of capital. Noting that the residents of affordable housing have the longest commuting distance among the other groups and a strong reliance on public transportation, promoting jobs-housing balance and improving the public transit system may greatly improve their situations.

3.6 Government Policy Is Vital for Individual Accessibility

The self-organisation process starts with the impacts of government policies and social norms on low-income workers, and then low-income workers use their individual decisions to transform the policies through the reproduction of practices. The process produces new behaviours that help low-income workers achieve their basic needs (Mote and Whitestone 2011). This book argues that unjust government policies are the root cause of accessibility problems in the process of urban development, and thus, government decision-makers who want to build a just society need to use social justice theories to guide their formulation of transport policies.

For example, the government of Hong Kong aimed to improve the mobility of drivers who commute between Hong Kong Island and the Kowloon Peninsula. It built three cross-harbour tunnels: the Cross-Harbour Tunnel, which connects Hung Hom in Kowloon and Causeway Bay in Hong Kong Island; the Eastern Harbour Crossing, which connects Cha Kwo Ling in Kowloon to Quarry Bay in Hong Kong Island; and the Western Harbour Tunnel, which connects West Kowloon and Sai Ying Pun (see Fig. 3.4). Under the build-operate-transfer (BOT) agreements between the government and the tunnels' private investors, the three tunnel operators can adjust the tunnel charges when their revenues fall short of the legally specified minimum estimation. The toll charges of the Western Cross Harbour Crossing are the most expensive in the city for traversing the harbour. As a result, many drivers feel that the toll charges in the Western Harbour Crossing are unaffordable and they choose to use the Hung Hom Cross-Harbour and Eastern Harbour Tunnels for routine work trips. The individual decisions and actions of these drivers result in heavy traffic congestion on the roads leading to the two tunnels, that is, the Hung Hom and Eastern Tunnels.

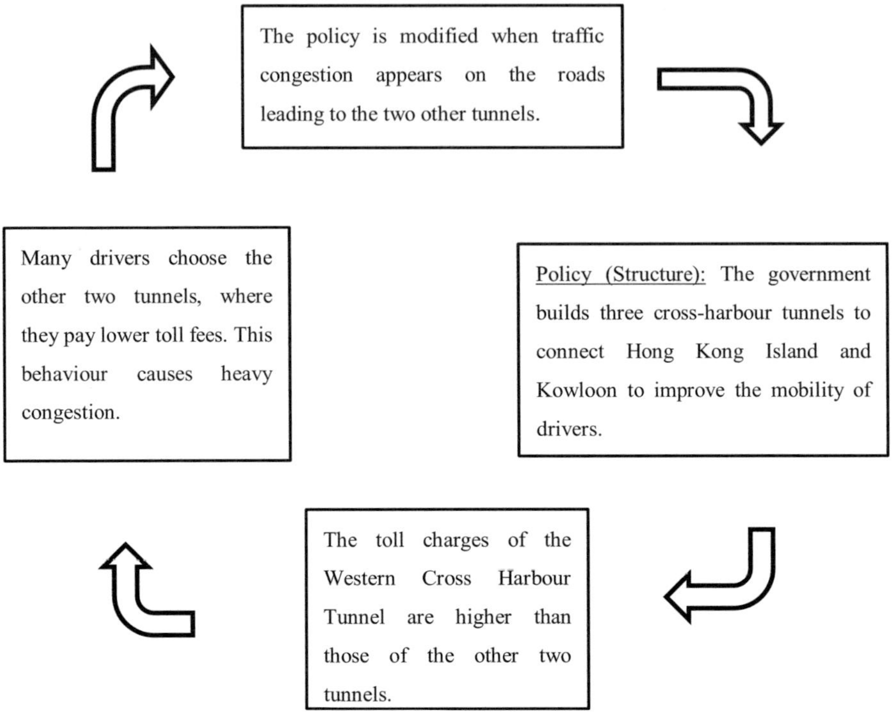

Fig. 3.3 Unjust transport policy causes congestion in Hong Kong

The self-organisation process involves drivers' individual decisions and the traffic policies on having built the cross-harbour tunnels by means of the BOT method, as seen with the Western Cross Harbour Tunnel. The financing and construction were the responsibility of a private enterprise, which was granted a concession to operate and collect tolls for 30 years. This recursive interaction between the BOT policy and the actions of the drivers has produced new travel behaviours, such as most drivers choosing to use the other two tunnels, which offer lower prices, and the policy causing traffic congestion (Fig. 3.3). This case shows that the BOT method is not a fair policy for improving the commutes of drivers and that it results in inaccessibility problems. In addition, the case study indicates that policy is a very important element in the self-organisation process. Therefore, in the next section, social justice theories will be introduced to help decision-makers understand and formulate fair policies to improve accessibility and social inclusion.

3.7 The Role of Social Justice in the Self-Organisation Process

This section aims to investigate the role of social justice in self-organisation and policy-making to improve accessibility and social inclusion. The self-organisation process is the recursive interaction between policy and individual actions to produce new structural properties that cause accessibility constraints and social exclusion problems. Structure and agency are interdependent variables because the structural properties of social systems are both the medium and the outcome of the practices they recursively organise (Giddens 1981, 1984). Government decision-makers who want to build a just society can follow social justice theories to modify unfair policies in the self-organisation process. This move would result in the production of new travel behaviours to resolve accessibility problems and minimise social exclusion. Hence, the role of social justice in the self-organisation process is important in urban transport planning, such as in formulating equitable suburbanisation policies that can provide a fair distribution of accessibility for displaced workers. This section tries to investigate the meanings of equality and justice and the principles of justice that are proposed in this book for amending the unjust policies in the self-organisation processes that take place in deprived neighbourhoods.

3.7.1 Self-Organisation Always Produces Efficient Policies but Ignores Equity in Urban Development

Self-organisation emphasises a mechanism through which complex social systems generate internal interactions among actors to produce new structural properties to cope with external influences. Social systems continuously evolve to find an optimum fit with their environment (Portugali 2000; Heylighen 2008). In urban transport planning, self-organisation is a spontaneously emerging order that delivers superior outcomes by nature and has much greater efficiency than do processes distorted by urban planners (Rauws et al. 2016).

For example, in Italy, city regeneration policies interact with the actions of retail workers to form self-organisation processes to promote retail-led regeneration (Brunetta and Caldarice 2014). The process involves all of the actors working in urban areas, including various stakeholders (institutions, civil society and business relations), helping construct efficient urban policies. The process also produces new structural properties, including business improvement districts, which are self-organised forms of urban management aimed at developing competitive strategies in view of the revitalisation of urban retail developments authorised by the government but promoted and managed by private operators.

The self-organisation process always produces efficient policies in the interaction between structure and individual actions so that the affected actors can achieve their intended goals. However, according to the results of the studies in Chaps. 4 and 5, the new structural properties produced by the self-organisation processes were arguably socially acceptable. Many of the affected low-income workers were experiencing social exclusion and accessibility problems.

For example, the self-organisation process sometimes produces Pareto efficiency, which is not in line with social justice principles. Pareto efficiency refers to a situation in which 'no possible improvement could raise the utility of one, without reducing the utility of any' (Sen 1992, p. 136) and is always the basis for government policy-making in allocating the optimal distribution of resources. For example, the Hong Kong SAR government implemented a policy of Pareto-improving reallocation (e.g., the possible improvement could raise the utility of at least one person without making any other individual worse off) that subsidises the travel of low-income workers without reducing the benefits to the public transport operators.

The relationship between Pareto efficiency and self-organisation provides evidence for the argument that self-organisation produces efficiency. Adam Smith (1937) proposed the idea of the 'invisible hand' as the self-organisation mechanism to organise the interactions among components in competitive market systems. Krugman (1996) illustrated the economic consequences of self-organisation: 'Self-organisation is something we observe and try to understand, not necessarily something we want' (Krugman 1996, pp. 5–6). Competitive markets produce Pareto-efficient outcomes, capturing the logic of Adam Smith's invisible hand. In addition, any Pareto-efficient outcome can be supported as a competitive market equilibrium (Mas-Colell et al. 1995). Given certain assumptions, the relationship can be proven mathematically.

The self-organisation process produces new structural properties that include Pareto efficiency, which ignores issues of social justice. For example, Pareto efficiency does not require an equitable distribution of wealth. One can imagine a society in which a small group of individuals controls a large share of resources and the remainder of society lives in poverty. This situation can be Pareto efficient despite the disparity between the rich and the poor (Grandy 2006). Government decision-makers who want to build a just society have the responsibility to ensure equitable institutional actions and decision-making processes about equitable resource distributions that promote people's quality of life. That is, all people in a society should have the freedom, choice and opportunity to move around freely and to access formal job opportunities, medical services, public education and housing. Hence, the role of social justice becomes very important in the self-organisation process that takes place during urban development.

3.7.2 Equality and Equity in the Distribution of Benefits and Burdens

The fair distribution of primary goods, including accessibility and the fulfilment of basic needs, is an important objective in the formulation of government policies.

Such a distribution is similar to the concept of social inclusion, which defines the development process of providing freedom, choice and opportunities for all people in order to achieve well-being and improve their access to the mainstream activities of society (World Bank 2018; Beyazit 2011).

In urban transport planning, the terms 'equity and justice' are often used interchangeably with the term 'equality', which can lead to confusion. Equality refers to people or groups having an equal distribution of resources and opportunities. For example, migrant workers and local *hukou* residents in urban China should have an equal distribution of mobility or the same level of public transit service regardless of their educational attainment, gender, abilities or income level. Therefore, as far as the objectives of public transit are concerned, it is impractical to achieve equality (Carleton and Porter 2018).

In Sweden, Solá et al. (2018) explore personal accessibility; they find that most planners agree on the planning goal of enabling the right of all citizens to have equal access to basic services:

> All residents of the studied municipalities should have equal access to basic services—a slogan version of the general Swedish welfare policy goal that all people have the right to access public services within an acceptable distance from home... Everyone should have access to a good life, and basic needs must be met first. (Solá et al. 2018; p. 7)

Therefore, rural migrant workers in urban China should have the right to equal access to employment and social services.

Justice can be understood as a broad moral and political ideal that relates to how accessibility, benefits and burdens are distributed in a society as well as to how rights and entitlements should be recognised and enforced (Pereira et al. 2017). Harvey (1973) defines social justice as "a just distribution" (1973, p. 98) and states that:

> Justice is essentially to be thought of as a principle for resolving conflicting claims...Social justice is a particular application of just principles to conflicts which arise out of the necessity for social cooperation in seeking individual advancement. The principle of social justice, therefore, applies to the division of benefits and the allocation of burdens. (p. 97)

Thus, justice is mainly concerned with just principles for particular resource distributions (e.g., equal rights to accessibility to employment, freedom to achieve basic needs). In addition, Miller (1976) states that:

> Principles of justice are distributive principles... The just state of affairs is that each individual has exactly those benefits and burdens which are due to him by virtue of his personal characteristics and circumstances (p. 20)

As far as transport planning is concerned, justice refers to a fair distribution of access to the means to achieve well-being. Rawls (1971) suggests that fair distribution means that even though inequalities in access take place in the process of social developments, the greatest benefit should be distributed to the disadvantaged. In other words, inequality in access to basic needs is considered fair if and only if government decision-makers provide the greatest levels of accessibility improvements to low-income workers during the process of urban development. For example, city governments invest huge financial resources in building public transit networks to cover all deprived districts so that low-income workers can easily take public transit services to their places of employment. In addition, Sen (1980, 1981, 1984, 1992) suggests that a just distribution of accessibility means that all people have the freedom, choice and opportunity to choose their transport mode, route, travel distance and trip frequency to satisfy their basic needs so that people can maintain their quality of life.

3.7.3 The Utilitarian Approach

According to the utilitarian principle, in an economically “just” society, policy-makers should assert that the best social policy is the one that gives the greatest total welfare to individual members of society, where total welfare is measured by summing utility numbers for all individuals. Bentham (Bentham 1988 [1776]) has advocated for this principle. According to the principle, an equitable distribution maximises the sum total of benefits irrespective of its individual distribution of resources, freedom and opportunities (Sen 1979). Thus, the moral judgement of a government’s policy is based exclusively on the policy’s consequences, particularly on how it maximises aggregate benefits, as seen with the *hukou* system in China.

Government transport policies can be evaluated by means of utility-based measures. For example, the cost and benefit analysis (CBA) in transport projects draws on the concept of the utility approach by comparing the expected costs and benefits of various alternative transport policy options. Costs and benefits are usually defined in monetary terms (Boisjoly and El-Geneidy 2017; Wee and Roeser 2013). The CBA has become the accepted standard for evaluating transport projects since the early 1960s (Nahmias-Biran et al. 2017).

To judge the well-being or satisfaction of a person exclusively in terms of utility has some obvious limitations with regard to government policy formulation. According to the concepts of the utilitarian approach and CBA, it is perfectly acceptable to override the individual rights of minorities if a transport policy promotes a greater good for a greater number of people (Pereira et al. 2017). This is deemed especially problematic in the case of transport system investment.

For example, in the Chief Executive’s policy address in Hong Kong in 2018, Mrs Lam Cheng Yuet-Ngor proposed drastically increasing the tolls of two cross-harbour government tunnels and lowering the tolls of one privately operated cross-harbour tunnel in order to relieve the tunnels’ congestion. The tolls for the Cross-Harbour Tunnel and Eastern Harbour Crossing for private cars would increase from HK$20 (US$2.6) to HK$40 (US$5.1). The government would subsidise the privately operated Western Harbour Tunnel, and its tolls would be reduced from HK$70 (US$9.0) to HK$50 (US$6.5) (see Fig. 3.4). The proposal aimed to achieve the re-distribution of traffic among the three tunnels to alleviate cross-harbour traffic congestion (SCMP, 11 October 2018).

The CE’s proposal could alleviate traffic congestion and air pollution to maximise the total benefits to Hong Kong society, but it is also an unfair policy proposal. A

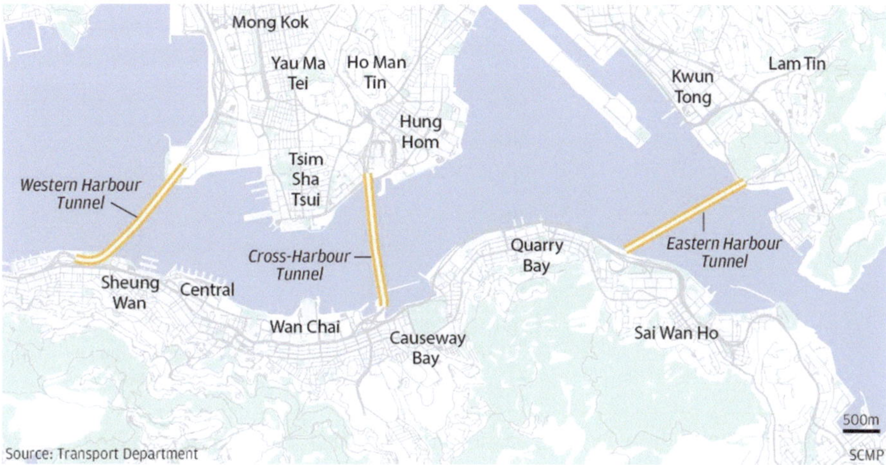

Fig. 3.4 Three cross-harbour tunnels in Hong Kong. Source: SCMP, 11 October 2018

large number of private car drivers would have to dig deeper into their pockets to pay the doubled toll. This would be a particularly heavy financial burden for low-income drivers who live in distant new towns and have to use one to three tunnels for a single daily work trip. Consequently, the proposal was scheduled to be tabled in the Legislative Council on March 2019 and was finally withdrawn by the government due to a lack of votes (Standard, 27 Mar 2019).

Furthermore, in Chaps. 4 and 6, this book investigated the self-organisation process in the inner suburbs of Guangzhou and Shanghai. The process focused on the interaction between the *hukou* system and the actions of rural migrants to produce new travel behaviours that caused social exclusion and accessibility problems. This section tries to evaluate the *hukou* system by means of the utilitarian approach. The *hukou* system was introduced in 1958 to deny citizens' movement between rural areas and cities. To provide adequate cheap labour to support economic growth, the system was relaxed after the economic reforms in 1978. In 2014, based on the productivity differences between agriculture and urban occupations in industry and services, every 1% increase in migration from rural to urban areas in China yields 1.2% more GDP (World Bank 2014). The *hukou* system allows rural *hukou* workers to move to cities to obtain jobs but prevents them from accessing social welfare and formal job markets.

The growth rate of China's GDP was 9.8% between 1995 and 2009 and it is forecasted to drop to 6.1% between 2016 and 2020. The *hukou* system denies migrants access to social services and formal job markets in host cities, which is the major factor in generating rapid GDP growth. Because of the *hukou* system, rural migrants' wages have remained low, which has made China the world's largest low-cost supplier of manufactured goods. Migrant workers became the backbone of the export industry and the manufacturing sector more generally in the 1990s. The

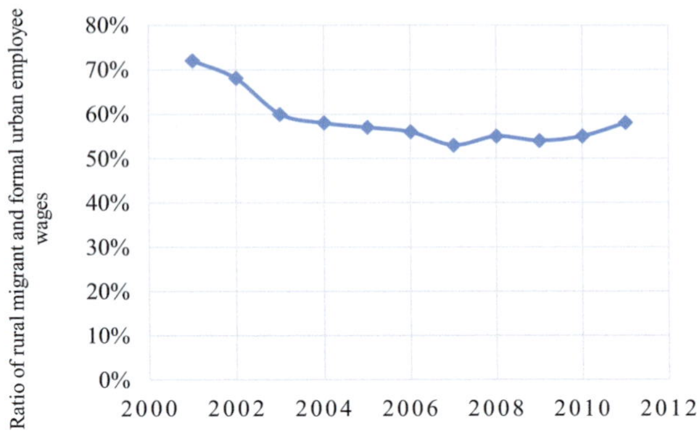

Fig. 3.5 Relative wages of rural migrants and formal urban employees, 2001–2011. Source: World Bank (2014)

rural migrant population in the coastal cities reached 132 million in 2006 (Chan 2009).

In 2014, it was hypothesized that if the host cities had extended social services to cover rural migrants in 2015, the cost would have been between 1.2% and 4.5% of China's GDP (World Bank 2014). Without having to pay the costs of covering migrants' social services, the host cities could spend more financial resources to maximise economic growth and industrial production in China.

Figure 3.5 depicts the relative wages of rural migrants and formal urban employees between 2001 and 2011. The wages of rural migrants stayed between 50% and 60% that of the local urban workers in this time period (World Bank 2014). Therefore, the cost and benefit analysis indicates that the *hukou* system is an efficient and competitive policy to help GDP growth in China. The utilitarian approach neglects the equitable distribution of income, opportunities, and the freedom of movement among social groups. However, the utilitarian approach is compatible with the ideology of communist countries such as China. The ultimate goal of a communist country is to establish a society in which a socioeconomic order is structured upon the common ownership of the means of production with free access for all people to the final goods. The working class, or all people in a communist society, owns everything, and there is no unfair distribution of resources. The individual's best interests are indistinguishable from society's best interests.

3.7.4 The Rawlsian Approach: A Just Distribution Must Give a High Priority to the Least Disadvantaged

The transport social justice literature has been developed further to a broader understanding of the concept in terms of primary goods, entitlement and capabilities (Pereira et al. 2017; Beyazit 2011; Rawls 1971, Sen 1992). This section investigates the Rawlsian approach that emphasises providing guidelines for policy-makers to consider the greatest benefit to the least disadvantaged and to aim for fairly distributing basic liberties and opportunities to all. For example, according to the Rawlsian approach, the policy-makers involved with the *hukou* policy in urban China should consider making it a high priority that rural migrants be provided with social welfare and formal job opportunities in their host cities (Rawls 1971).

Rawls argues that the focus should not be on maximising the total benefits to society but rather on the distribution of certain kinds of benefits, such as basic liberties, freedom of movement and income, which, according to him, are 'primary' for all persons. Primary social goods according to Rawls (1971) are as follows:

- Basic liberties, including the freedom of association;
- Freedom of movement (including accessibility to needed activities) and choice of occupation;
- Powers and prerogatives of offices and positions of responsibility;
- Income and wealth; and
- The social bases of self-respect.

Rawls states that the freedom of movement, accessibility to needed activities and choice of employment must be complete and have priority over the distribution of other primary social goods (Rawls 1971). Hence, Rawls' list of social goods or human rights should be considered in government policy-making, and his approach is considered fairer than the utility approach. In accordance with Rawls' theory, low-income workers in urban China should have the freedom to choose their employment and have choices in moving about to find jobs. Thus, the *hukou* system is an unjust policy.

According to Rawls (1971), the two principles of justice are as follows:

> First Principle: Each person is to have an equal right to the most extensive total system of equal basic liberties compatible with a similar system of liberty for all.
>
> Second Principle: Social and economic inequalities are to be arranged so that they are both: (a) to the greatest benefit of the least advantaged, consistent with the just savings principle, and (b) attached to offices and positions open to all under conditions of fair equality of opportunity. (p. 266).

The first principle is often called the greatest equal liberty principle. The second, until point (b), is known as the difference principle. The above two principles are used to govern the assignment of rights and duties and to regulate the distribution of social and economic advantages for a government in its urban transport planning. The first principle argues that every person in a just society should have an equal right to the most extensive basic liberties, which include the freedom of movement and accessibility to the means to fulfil basic needs such as education, jobs and medical services. The second principle maintains that some inequality of wealth and income is probably necessary in order to maintain high levels of productivity. Any unequal distribution of wealth and income should be such that those who are worst off are better off than they would be otherwise. People who are better off in society should help the disadvantaged to achieve their basic freedom and rights, which include the freedom to access opportunities.

The ideas behind these two principles can be expressed as follows: the basic structure of a just society ensures the distribution of certain primary goods, such as liberty and opportunity, income and wealth, and it is supposed that in the initial arrangement, all the primary social goods are equally distributed. The later socio-economic and political development planning give rise to inequalities in wealth and opportunity in the process of urban development. If government decision-makers place high priority on helping low-income people achieve their primary social goods in the process of policy formulation, these policies are considered fair or in accordance with the principles of justice (Pereira et al. 2017).

For example, Chap. 6 investigates the suburbanisation of Shanghai. The municipal government used land finance policy to generate an extensive land sales in order to finance the construction of rail transit networks to connect the new towns and the city centre. At the same time, many displaced low-income workers in the new towns chose to take the rail transit services to new job opportunities in the suburbs and city centre. The interaction between the land finance policy and individual actions of workers produced new travel behaviours, which included many displaced workers commuting to their places of employment by using fast rail transit services.

The land finance policy adopted by the municipal government is consistent with the difference principle that inequalities in the distribution of accessibility to employment are permissible if and only if they are to the greatest benefit of the disadvantaged (Rawls 1971). The policy promotes the accessibility of low-income displaced workers without reducing the benefits to private car users.

However, the Rawlsian approach runs into the problem of interpersonal comparisons, such as people differing in their relative abilities, educational achievements and social roles and in their experience with the policies that are created by the difference principle (Nahmias-Biran et al. 2017). For example, in the above case, the displaced workers in the new towns of Shanghai adopted rail transit services that could satisfy their daily travel needs, such as having access to employment, recreation, education and social events. It seems, however, that goods possession is not a fair way of assessing accessibility. This is because even if unskilled displaced workers could take the rail transit services to work, they could not find new jobs in the suburban areas. Since their low educational achievements and low level of skills constrained them in finding new jobs, their level of accessibility was lower than that of the rural migrant workers who were able to ride bicycles to find informal jobs in the suburbs.

Similarly, the Shanghai municipal government built trail transit networks for displaced workers. However, decision-makers may still need to consider whether these disadvantaged people actually have access to rail transit networks, i.e., whether the rail transit network covers all of the dispersed locations in the new towns that are the result of urban sprawl. In the following section, this chapter introduces the capability approach, which suggests that the way a just government should deliver a just policy is to ensure that all people have the freedom, choice and opportunity to achieve basic functionings in order to maintain their quality of life.

3.7.5 The Capability Approach

Sen (1980, 1981, 1984, 1992) proposed a justice principle for resource distribution called the capability approach. According to the Capability Approach, the operationalised capabilities are referred to as the individual's functionings (e.g., the individual's realised behaviour) (Ryan et al. 2019). This approach entails two core claims for the equitable distribution of human rights to maintain people's quality of life in society: first, it is a human right to have the freedom to achieve functionings, such as having access to employment and housing. The second claim is that the freedom to achieve well-being is to be understood in terms of people's capabilities. For example, people should have the real ability to obtain the opportunities, freedom, and choice needed to fulfil their basic needs, such as being educated, employed, housed in a decent home and suitable living environment, and part of a supportive social network.

The theory focuses directly on the influence of structures or government policies on providing choices for and facilitating the actions of people in fulfilling their basic needs. Sen (1980, 1981, 1984, 1992) indicates that equitable distribution should focus on capabilities rather than on resources such as transport systems and services, as upheld by Rawls. Thus, a government characterised by justice and equity should ensure the equitable distribution of freedom, opportunity and choice (e.g., commuting to employment and being employed) among all people rather than focusing on providing the equal distribution of public transport services.

City governments should ensure that transport and other policies facilitate all people having the freedom, opportunity and choice to use their own labour power in change for wages in the labour market. For example, in Hong Kong SAR, a low-income working mother who lives a distant new town should have the choice to seek jobs near her home or in urban areas and to take buses or the Mass Transit Railway to reach her place of employment.

In addition, the capability of low-income workers in urban China represents a degree of freedom to freely choose between different retail, construction and manufacturing jobs in formal labour markets. Concepts such as freedom, choice and the availability of opportunities become crucial (Sen 1999). In other words, a rural migrant in urban China should have the opportunity to make individual decisions and take individual actions to choose to work in a TVE in the inner suburbs or to work in a state-owned enterprise in urban China (Beyazit 2011).

In light of the self-organisation process that is taking place in deprived urban neighbourhoods in China, decision-makers who want to build a just society should follow the capability approach and transform the *hukou* system so that rural migrants in urban China can have the freedom and choice to choose their jobs, housing and medical services in their host cities. The rural migrants should have the freedom to choose between various transport modes, travel times, and travel distances to reach various jobs without structural constraints. Furthermore, Chap. 5, which investigates Xiguan, the inner city in Guangzhou, describes how many low-income workers in the deprived inner city feel accessibility constraints because of narrow roads and the poor provision of public transit services. Hence, the formulation of policies based on the capability approach in urban redevelopment and urban revitalisation could improve the accessibility problems of the residents in Xiguan.

Sen himself has pointed out that capability deprivation is a form of social exclusion (Sen 2000). Rural migrant workers live in poverty not because of a lack of employment opportunities for rural migrants in urban China but because the *hukou* system excludes them from accessing formal employment opportunities. In Chap. 4 of this book, rural migrants' social exclusion problems are investigated through the self-organisation approach. Several kinds of social arrangements, such as education, employment, social relations, housing and the gendered division of household labour, are very important aspects in determining migrants' well-being in urban China and hence in relating to their experience of social exclusion.

3.8 Self-Organisation Process and Unjust Policies

In the face of unjust social policies, the self-organisation process in social systems can help to transform inequitable social policies through the actions of individuals. The process is based on the concept that the mechanism of the self-organisation approach is presented by the cycle of structure and agency (Giddens 1981, 1984). For example, in Chap. 5, the results of a case study on Xiguan, a historic and outdated inner city in Guangzhou, indicate that when an unfair government redevelopment policy or structure influences the culture and norms of the local residents, the residents will carry out self-organised actions to transform the unjust policy. According to the author's interview results, the Guangzhou municipal government originally adopted a redevelopment policy that followed privately funded and property-led redevelopment approaches to demolish the old buildings, resettle the affected residents and build modern new constructions, as was the case with the Xintiandi redevelopment project in Shanghai (He and Wu 2005), which required tearing down all of the historical architecture in Xiguan.

The residents developed bottom-up emergence by carrying out a demolition review during the early conceptual stages of the plan and finally transformed the unfair policy; the government abandoned the original idea of demolition and adopted the new plans to conserve the historic buildings. The Guangzhou government reclaimed the responsibility for inner-city redevelopment and excluded the private sector from upcoming redevelopment projects. This example suggests that when a policy is unfair, people will make individual decisions and take individual actions to transform it.

3.9 Conclusion

Chapter 3 reviews the concepts of accessibility, travel needs, the space-time prism and social justice theories. Structural factors, e.g., the built environment, alone cannot explain complex travel behaviours. Transport planners should not ignore the importance of individual decisions and actions on travel behaviours and accessibility. This is because new travel behaviours are products of the self-organisation process, which is the interaction between structure and travel practices (Giddens 1984). Accessibility is defined as the ability of an individual to penetrate barriers to accessing the mainstream activities of their society. The definition involves two important concepts, the term 'ability' is linked to the right of the low-income workers in urban China to move freely to meet basic needs, and the term 'mainstream activities of society' is linked to the social exclusion of low-income workers. Accessibility means that low-income workers have freedom, choice and opportunities in fulfilling basic needs and participating in the mainstream activities of society. Since the self-organisation approach involves the interaction between government policies and individual actions, decision-makers can improve the accessibility and reduce the social exclusion of low-income workers by changing either the policy or individual actions in the self-organisation process. This book suggests that government policy is the main tool for influencing personal accessibility.

This chapter also reviews travel needs and their relationship to the self-organisation process in urban development. Travel needs refer to the needs of low-income workers to use transport tools to reach the means to fulfil their basic needs. Chapter 6, which is about the suburbanisation of poverty, discusses how many displaced workers in Shanghai have faced unmet travel needs as a result of urban sprawl and the inadequate provision of rail transit services. The current chapter also investigates the impacts of a person's individual decisions on commuting to obligatory and discretionary activities. The results of the studies in Chap. 4 indicate that many rural migrant workers spend most of their time intervals on obligatory activities, resulting in accessibility problems. Most low-income workers allocate more time intervals to employment and fewer time intervals to social and recreational activities. Furthermore, the concept of the travel time ratio is used to investigate the impacts of individual decisions on the allocation of travel time and activity time for needed activities.

The new travel behaviours and the related accessibility problems that are produced by the self-organisation process can be measured by the space-time prism approach, as seen in Chaps. 4 and 5, which show how low-income workers were found to take short-distance trips and use slow transport modes to reach activities. The incorporation of the self-organisation approach into the space-time prism approach will help transport planners understand that the short-distance trips of workers lead to the smaller physical size of their PPAs, and thus the workers experience accessibility problems in accessing non-work activities.

The second part of this chapter reviews the role of social justice in the self-organisation process. The self-organisation process always produces efficient policies but ignores equity or justice. The investigations on the relationship between Pareto efficiency and competitive market systems (which are self-organisation processes) confirm this argument. Many government policies that seek Pareto efficiency cannot distribute resources using just principles. The investigation indicates that justice is important for policy formulation in the process of urban development. Government policy-makers who want to build a just society should understand and transform unjust policies in self-organisation processes into just policies so as to improve the accessibility and social inclusion of the poor.

Justice theories focus on a fair distribution of opportunities, responsibilities and primary goods. The utilitarian approach suggests maximising the sum total of all benefits irrespective of their individual distribution. The Rawlsian approach proposes that social and economic inequalities be arranged so that they are to the greatest benefit to the least advantaged. This approach aims to fairly distribute resources and opportunities to all. The capability approach emphasises the equitable distribution of freedom, choice and opportunities for all individual members in society so that they can achieve well-being. As far as the self-organisation approach is concerned, this book proposes that the capability approach is the best approach for improving the social inclusion of low-income workers in urban China. If the Chinese government cadres who want to build a just society adjust their unjust policies, such as the *hukou* system, the rural migrants may be able to achieve freedom, opportunities and choice in fulfilling their basic needs.

References

Ben-Akiva M, Lerman SR (1979) Disaggregate travel and mobility-choice models and measures of accessibility. In: Hensher DA, Stopher PR (eds) Behavioural travel modeling. Croom-Helm, London, pp 654–679

Bentham J (1988 [1776]) A fragment on government. Cambridge University Press, Cambridge

Beyazit E (2011) Evaluating social justice in transport: lessons to be learned from the capability approach. Transp Rev 31(1):117–134

Boisjoly G, El-Geneidy AM (2017) The insider: a planners' perspective on accessibility. J Transp Geogr 64:33–43

Brunetta G, Caldarice O (2014) Self-organisation and retail-led regeneration: a new territorial governance within the Italian context. Local Econ 29(4–5):334–344

Cao M, Hickman R (2019) Understanding travel and differential capabilities and functionings in Beijing. Transp Policy 83:46–56

Carleton PR, Porter JD (2018) A comparative analysis of the challenges in measuring transit equity: definitions, interpretations, and limitations. J Transp Geogr 72:64–75

Chan KW (2009) The Chinese *Hukou* system 50. Eurasian Geogr Econ 50(2):197–221

Chen C, Mokhtarian PL (2006) Trade-offs between time allocations to maintenance activities/travel and discretionary activities/travel. Transportation 33:223–240

China Daily (2017) High-speed trains to reduce travel time between west China cities, December. http://usa.chinadaily.com.cn/china/2017-12/05/content_35217290.htm. Accessed 2 Feb 2018

Dijst M, Vidakovic V (2000) Travel time ratio: the key factor of spatial reach. Transportation 27:179–199

Geurs KT, Wee BV (2004) Accessibility evaluation of land-use and transport strategies: review and research directions. Transp Geogr 12:127–140
Giddens A (1979) Central problems in social theory: action, structure and contradiction in social analysis. University of California Press, Berkeley
Giddens A (1981) A contemporary critique of historical materialism. Vol. 1. Power, property and the state. Macmillan, London
Giddens A (1984) The constitution of society. Polity Press, Cambridge
Grandy C (2006) Through a glass, darkly: an economic view of fairness, globalization, and states. In: Jim D, Dick P, Yongseok S (eds) Fairness, globalization, and public institutions: East Asia and beyond. University of Hawai'i Press, Honolulu, pp 49–60
Hägerstrand T (1970) What about people in regional science? Reg Sci Assoc Pap 24(1):7–21
Hansen WG (1959) How accessibility shapes land use. J Am Inst Plann 25:73–76
Harvey D (1973) Social justice and the city. Basil Blackwell, Oxford
He S, Wu F (2005) Property-led redevelopment in post-reform China: a case study of Xintiandi redevelopment project in Shanghai. J Urban Aff 27(1):1–23
Heylighen F (2008) Complexity and self-organization. In: Bates MJ, Maack MN (eds) Encyclopedia of library and information sciences, 3rd edn. Taylor & Francis, London, pp 1215–1224
Kim HM, Kwan MP (2003) Space-time accessibility measures: a geo-computational algorithm with a focus on the feasible opportunity set and possible activity duration. J Geogr Syst 5(1):71–91
Krugman PR (1996) What economists can learn from evolutionary theorists, (transcript of speech given to the European Association for Evolutionary Political Economy, Rome). http://web.mit.edu.eproxy1.lib.hku.hk/krugman/www/evolute.html
Levitas R, Pantazis C, Fahmy E, Gordon D, Lloyd E, Patsios D (2007) The multi- dimensional analysis of social exclusion, department of sociology and school for social policy, Townsend centre for the international study of poverty and Bristol institute for public affairs. University of Bristol, Bristol
Mao Z, Etterma D, Dijst M (2018) Analysis of travel time and mode choice shift for non-work stops in commuting: case study of Beijing, China. Transportation 45:751–766
Mas-Colell A, Whinston MD, Green JR (1995) Chapter 16: equilibrium and its basic welfare properties, microeconomic theory. Oxford University Press, Oxford
Meyer D, Miller EJ (2001) Urban transport planning: a decision-oriented approach. McGraw-Hill, New York
Miller D (1976) Social justice. Claredon Press, Oxford
Miller HJ (1991) Modelling accessibility using space-time prism concepts within geographical information systems. Int J Geogr Inf Syst 5(3):287–301
Miller EJ (2018) Accessibility: measurement and application in transportation planning. Transp Rev 38(5):551–555
Mote JE, Whitestone Y (2011) The social context of informal commuting: slugs, strangers and structuration. Transp Res A 45:258–268
Nahmias-Biran BH, Martens K, Shiftan Y (2017) Integrating equity in transportation project assessment: a philosophical exploration and its practical implications. Transp Rev 37 (2):192–210
Neutens T, Witlox F, Demaeyer P (2007) Individual accessibility and travel possibilities: a literature review on time geography. Eur J Transp Infrastruct Res 7(4):335–352
Patterson Z, Farber S (2015) Potential path areas and activity spaces in application: a review. Transp Rev 35(6):679–700
Pereira RHM, Schwanen T, Banister D (2017) Distributive justice and equity in transportation. Transp Rev 37(2):170–191
Portugali J (2000) Self-organization and the city. Springer, Heidelberg
Pyrialakou VD, Gkritza K, Fricker JD (2016) Accessibility, mobility, and realized travel behavior: assessing transport disadvantage from a policy perspective. J Transp Geogr 51:252–269
Rauws W, De Roo G, Zhang S (2016) Self-organisation and spatial planning: an editorial introduction. Town Plan Rev 87(3):241–251

Rawls J (1971) A theory of justice. Oxford University Press, Oxford
Ryan J, Wretstrand A, Schmidt SM (2019) Disparities in mobility among older people: findings from a capability-based travel survey. Transp Policy 79:177–192
Sen A (1979) Personal utilities and public judgments: or what's wrong with welfare economics? Econ J 89:537–558
Sen A (1980) Equality of what? Choice, welfare and measurement (Cambridge, MA: Harvard University Press), 1997 [Originally published in The Tanner Lectures on Human Values, Vol. I, by the University of Utah Press and Cambridge University Press, 1980]
Sen A (1981) Poverty and famines: an essay on entitlement and deprivation. Clarendon Press, Oxford
Sen A (1984) Resources, values and development. Oxford, Basil Blackwell
Sen A (1992) Inequality reexamined. Oxford University Press, Oxford
Sen A (1999) Development as freedom. Oxford University Press, Oxford
Sen A (2000) Social exclusion: concept, application and scrutiny. Social development papers no 1. Office of Environment and Social Development, Asian Development Bank, Manila
Smith A (1937) An inquiry into the nature and causes of the wealth of nations [1776]. In: Cannan E (ed) . Random House, New York
Smith N, Hirsch D, Davis A (2012) Accessibility and capability: the minimum transport needs and costs of rural households. J Transp Geogr 21:93–101
Solá AG, Vilhelmson B, Larsson A (2018) Understanding sustainable accessibility in urban planning: themes of consensus, themes of tension. J Transp Geogr 70:1–10
Van Wee B, Geurs K (2011) Discussing equity and social exclusion in accessibility evaluations. Journal of Emerging Technologies and Innovative Research 11(4):350–367
Wang D, Chai Y (2009) The jobs–housing relationship and commuting in Beijing, China: the legacy of Danwei. J Transp Geogr 17:30–38
Wang D, Zhou M (2017) The built environment and travel behavior in urban China: a literature review. Transp Res D 52:574–585
Wee BV, Roeser S (2013) Ethical theories and the cost–benefit analysis-based *Ex Ante* evaluation of transport policies and plans. Transp Rev 33(6):743–760
World Bank (2014) *Urban China:* toward efficient, inclusive, and sustainable urbanization, development research center of the state council, the People's Republic of China. World Bank Group, Washington, DC
World Bank (2018) Social inclusion. http://www.worldbank.org/en/topic/social-inclusion. Accessed 20 Mar 2018
Yang W, Chen BY, Cao X, Li T, Li P (2017) The spatial characteristics and influencing factors of modal accessibility gaps: a case study for Guangzhou, China. J Transp Geogr 60:21–32
Zacharias J (2012) Resisting motorization in Guangzhou. Habitat Int 36:93–100
Zhang X, Wang J, Kwan MP, Chai Y (2019) Reside nearby, behave apart? Activity-space-based segregation among residents of various types of housing in Beijing, China. Cities 88:166–180
Zhao J, Benthlage M, Thierstein A (2017) Residence, workplace and commute: interrelated spatial choices of knowledge workers in the metropolitan region of Munich. J Transp Geogr 62:197–212

Chapter 4
Self-Organisation Processes Taking Place in a Migrant Enclave in Guangzhou

4.1 Introduction

Guangzhou, the capital city of Guangdong province in China, was chosen for this empirical study because it is one of the main cities in the Pearl River Delta (PRD) economic and manufacturing zone. The PRD is known for its pioneering role in China's open-door policy and is one of the largest suppliers of manufactured goods in the world. Some of the leading products include footwear, lighting fixtures, furniture and toys (HKTDC 2016). Guangzhou has witnessed rapid urbanisation and faced a large influx of migrants from rural provinces seeking manufacturing processing jobs. In 2017, the physical size of Guangzhou was 7434 km^2, with a permanent population of 14.50 million; the rural migrant population accounted for approximately 38% of the total permanent population, and most of them lived in the urban villages (Guangzhou Statistics Bureau 2018).

The *hukou* system, which is an institution controlling population movement in China, prevents rural migrants from accessing social welfare in urban China. The migrants are excluded from accessing formal jobs, public housing, education and medical services. Migrants move to urban villages to find industrial jobs. There were 138 urban villages in 2010 (Liu et al. 2010). This study focuses on Qingfengcun, an urban village where the standard of living is low and there are dense buildings, packed inner streets, and poor hygiene conditions. The coordination and mutual adaptation among migrants and local villagers eventually started to turn the study urban village into an industrial park.

Few transport studies consider how life changes and self-organisation processes influence individual travel behaviours. To fill this research gap, this chapter investigates the influence of the self-organisation process that is taking place in the study urban village and how the process is producing new travel behaviours that lead to social exclusion and accessibility problems. Self-organisation is a phenomenon in

J. C.-y. Lau, *Self-Organisation Shapes Travel Behaviours and Social Exclusion in Deprived Urban Neighbourhoods of China*,
https://doi.org/10.1007/978-981-15-2252-9_4

which social norms and government policies interact with individual actions to produce new behaviours and social changes (Portugali 2000; Giddens 1984).

The travel behaviours of migrants in the study urban village are difficult to measure because travel practices are the result of complex interactions between the individual actions and various levels of government rules and social norms in the study urban village. To determine the new travel behaviours of migrants, this study used in-depth interviews to explore the travel patterns of migrants in urban villages. A travel survey was also carried out to find the observable new travel patterns of migrants. The self-organisation process that took place in the urban village produced new structural properties, which included migrants' short-distance travel behaviours to jobs and social activities. The migrants used slow transport modes to reach activities. The new travel behaviours led to social exclusion and accessibility issues.

Social exclusion is defined as a multi-dimensional phenomenon that deprives low-income workers' of access to the means to achieve well-being due to resource deprivation and the lack of freedom, choice and capability to participate in the mainstream activities of society (United Nations 2016; Sen 1984, 2000). In the study urban village, the variables that result in social exclusion include personal income, employment status, the gendered division of household labour, the *hukou* and land-use systems, social networks, and social discrimination. These variables interact with the individual actions of migrant workers and local villagers to produce new structural properties that cause social exclusion problems, such as the migrants' work trips being spatially restricted to the enclave. These travel patterns result in the exclusion of migrant workers from accessing the mainstream activities of society (Fuchs 2003).

In addition, the self-organisation process involves the local villagers in the study urban village producing new structural properties that include the development of township and village enterprises (TVEs) and affordable housing. The social structural factors include the land acquisition policy and the dualistic land-use system. Rural land is collectively owned, while urban land is state owned. However, the Guangzhou municipal government has acquired the rural land of the study urban village for urban use. To save costs in land acquisition, the original site of the study urban village has not been acquired by the state for development projects and thus remains collectively owned. Without farmland to grow crops, the local villagers build housing and factories on their collectively owned land to earn income to maintain their quality of life. The actions of villagers and migrants interact with policies that also produce jobs-housing co-location processes to cut travel times. Chapter 4 hypothesises that the self-organisation process, which is taking place in Qingfengcun, the study urban village, produces new structural properties and accessibility and social exclusion problems.

In this chapter, the author poses the following four research questions: (1) How does the self-organisation process produce new travel behaviours that result in the accessibility problems of migrants in reaching non-work activities? (2) How does the self-organisation process produce new travel behaviours that lead to a jobs-housing co-location process? (3) How does the self-organisation process produce new structural properties that result in the exclusion of migrants and local villagers from participating in the mainstream activities of society? (4) What should the policy-makers who want to build a just society do to minimise social exclusion? In addition, the author expects that the investigation of the above four research questions can shed light on the research on social exclusion, urban transport planning and the self-organisation approach.

The remainder of this chapter is structured as follows. The next section reviews the literature on the cycle of structure and agency in urban villages, which leads to social exclusion and accessibility problems for rural migrants. This is followed by the presentation of the results of in-depth interviews and a travel survey in Qingfengcun, the study village in Guangzhou. The presentation mainly focuses on the individual actions of the rural migrants and how the local villagers interact with the structures in the self-organisation processes. The final two sections in this chapter analyse the travel data of the respondents, determine their new travel behaviours and discuss methods to improve their accessibility to distant non-work activities and minimise their social exclusion problems.

4.2 Self-Organisation Process Influences the Routine Activities of Actors

According to Giddens (1981, 1984), the structure consists of rules and resources that are recursively implicated in the reproduction of social systems and are stored within the memory traces of individuals. The self-organisation process in urban villages starts with rules and resources that constrain and enable rural migrant workers' actions in their routine activities. The second stage of the process involves rural migrant workers reproducing and transforming those rules and resources to find jobs and borrow start-up cash for developing small retail businesses in urban villages (see Giddens 1984).

The term 'rules' refers to the rules of social life, such as government policy, culture and social norms, while the term 'resources' refers to anything that can serve as a source of power in social interactions, such as social networks. The self-organisation process also includes the social exclusion process (e.g., low-income workers interacting with economic reforms), which in turn includes structures that interact with the individual actions to produce new structural properties that exclude people from participating in the mainstream activities of society (Stanely et al. 2019; Schwanen et al. 2015; Stanley and Vella-Brodrick 2009). For example, in the investigation of urban villages, the structures that are related to the social exclusion of migrants can be divided into the following types: First, the impacts of social norms, such as the gendered division of household labour that constrains working mothers in finding distant employment. Second, the social networks that enable migrants to concentrate in urban villages to find jobs and hinder migrants in seeking social contact with other local residents. Third, systems such as the *hukou* system, which prevents migrants from finding formal jobs and accessing social welfare. In addition, the social exclusion process in urban villages involves different types of individual decisions and actions from rural migrants, villagers and village collectives. The actions and decisions of these actors transform the structures as they seek to use their land and labour power in exchange for income and food.

4.3 The *Hukou* System Excludes Rural Migrants from Accessing Formal Jobs and Social Welfare

In 1949, the Chinese Communist Party took control of China, which was a highly agricultural nation. Approximately 89% of its citizens lived in rural areas. However, as efforts to industrialise increased, more and more rural residents flocked to cities in search of better economic opportunities. By 1954, rural and urban citizens had been registered with the state, and rigorous regulations on the conversion of *hukou* status had been implemented. In 1958, the People's Republic of China *Hukou* Registration Regulation was signed into law, and it remains fully in force today. This law divided the population into those with rural *hukou* status and those with urban *hukou* status and grouped all citizens by locality. Because the central government prioritised industrialisation, state welfare programmes, which were tied to *hukou* status, heavily favoured urban residents, those holding rural *hukou* status were unable to access these benefits and were saddled with inferior welfare policies.

The *hukou* system required all internal migration to be subject to approval from the authorities at the destination. The *hukou* classification of children would follow that of the mother (Chan 2009). In 1978, the government lifted the restrictions on rural residents migrating to cities in order to provide cheap labour for economic reforms. However, the *hukou* system still did not allow rural migrants to change their residence or claim any benefits in the cities. In this book, the term rural migrant workers refers to those workers with a rural *hukou* status who work and live in host cities. These individuals are not necessarily from rural areas. Many of them were even born in urban areas. Because of the inflexibility of the *hukou* system, they remain classified as rural migrants, as is the case for those workers who were born in Guangzhou, work in Shanghai, and are still classified as rural *hukou* workers in Shanghai.

Furthermore, rural migrants have been labelled a 'floating population' in urban China, suggesting that their residence in the major cities is only temporary and that they are unlikely to develop strong local ties. Most of them return to their rural towns every year during the period of the Spring Festival. The total number of rural migrant workers in China's cities reached a new high of 287 million in 2017. Today, according to the public housing rules, rural migrants cannot rent or purchase municipal or work-unit public housing directly because only sitting tenants (local urban residents) can do so. All public housing types, including economic and comfortable housing and affordable rental units, are reserved for only local *hukou* residents in cities. Migrants are allowed to purchase commodity housing in the private sector, but this path is unaffordable for most migrants (Wu 2006).

In the face of the *hukou* system, many rural migrants choose to find informal jobs in urban villages, which are located in the inner suburbs of host cities. The migrants face capability deprivation; they do not have the opportunity, freedom or choice to use their own labour power to trade for income, decent homes and suitable living environments (Zhu 2016; Sen 2000). The urban villages are spatially segregated from the main residential neighbourhoods in host cities. Thus, the *hukou* system interacts with the actions of migrants to produce new structural properties that lead to social exclusion and accessibility problems.

4.3.1 The **Hukou** *System and Occupational Segregation*

Great segmentation between opportunities has been found between migrant workers and local residents in urban China; this is exemplified by rural *hukou* migrants who are excluded from accessing professional, managerial and governmental jobs. Municipal governments set rules that constrain employers from hiring migrant workers in specific occupations in order to protect local workers. Many high-income commercial job opportunities under the management of local authorities are also not accessible to migrants. These commercial job opportunities include cashiers, managers and administrative public transport staff as well as office jobs related to the financial and insurance sectors and accountancy. As a result, the chance of migrants finding employment in city centres is low (Fig. 4.1).

Occupational segregation is most severe in government institutions and least severe in the private sector, leading to earnings disparities between rural migrants and local urban workers in different work unit sectors (Zhang and Wu 2017; Zhao and Howden-Chapman 2010). As a result, most rural migrants in urban areas find low-end jobs in different industry sectors, such as service industries such as transport, logistics, manufacturing and construction. Moreover, even within the same occupation, migrant workers suffer serious wage discrimination. A large wage gap exists between rural migrants and urban workers (Liu 2015). The positions open to migrants often have long working hours and involve dirty, dangerous and challenging work, particularly for those who work in the construction, manufacturing, sales or catering industry sectors. Thus, the *hukou* system interacts with the actions of migrants to produce new structural properties that prevent them from accessing formal jobs and produce occupational segregation in urban China.

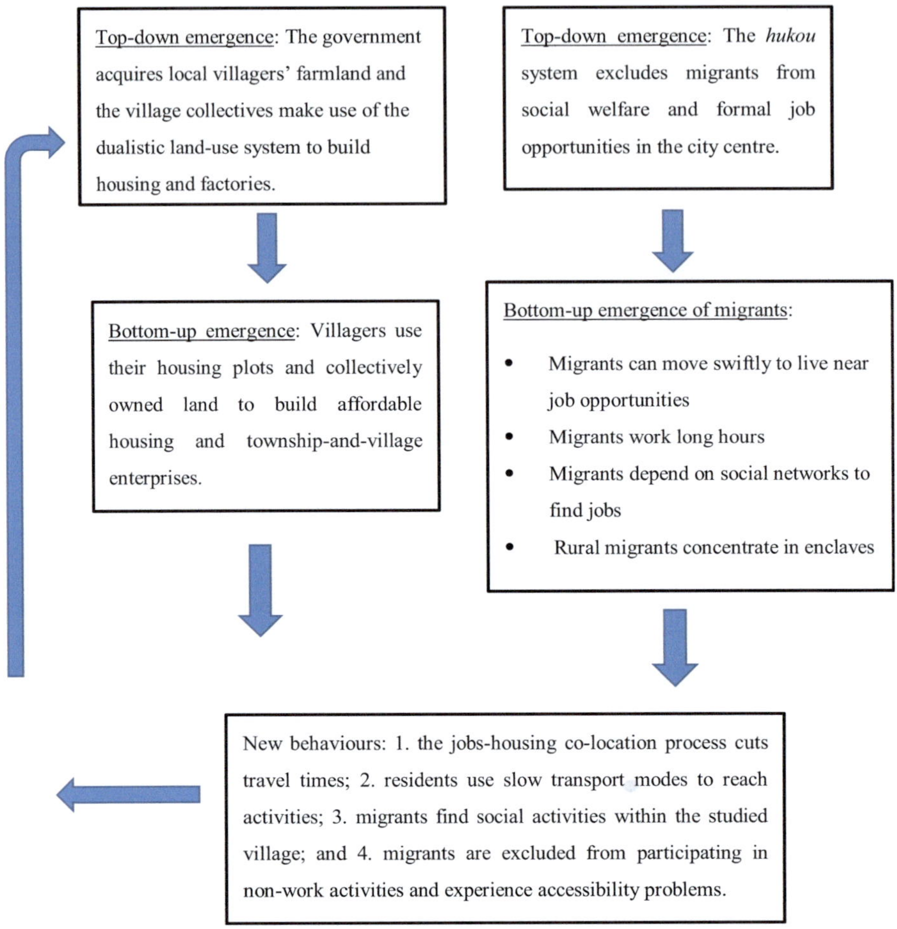

Fig. 4.1 Self-organisation processes in urban villages produce new behaviours

4.3.2 Rural Land Acquisition and Individual Actions of Local Villagers

In China, there are two tracks in the system of land ownership in which urban land is owned by the state and village collectives own rural land. Urban land-use rights can be transferred to urban land markets, and rural land cannot be sold. In the process of urban expansion through compulsory rural land acquisition, villagers' farmland, most of which is located in the inner suburbs, is acquired by the municipal governments to facilitate urban-based economic growth. During the process of land acquisition, the built-up areas in the villages are left untouched (Huang et al. 2017). The municipal governments usually rezone the acquired farmland to establish economic and technological development zones (ETDZs), and other acquired land is put to land auction. As a result, these landless farmers cannot use their labour power and farmland to grow crops in exchange for income, resulting in entitlement failure. Rural protests over these land grabs have constituted the largest source of mass disturbances in China in recent years (Chung and Unger 2013; He et al. 2009).

In the face of entitlement failure, villagers use their decisions and actions to transform the rural land acquisition policy through the production and reproduction of routine activities. As compensation for the loss of their land, each household can receive a one-off lump sum payment from the municipal government; they can then use the compensation payments to develop rental-housing businesses. In addition, village collectives often use the compensation payments and take out bank loans to convert built-up land in villages into TVEs, leasing the factory buildings to local and outside enterprises. In urban villages, village collectives distribute housing plots (Zhaijidi) to individual households. Most of the households use their housing plots to build private multi-storey rental buildings (Huang et al. 2017; He et al. 2009). The urban villages are gradually transformed into manufacturing production centres and migrant enclaves. In 2013, the land in the urban villages was no longer owned by rural collectives but rather by shareholding companies. The collectively owned land was transferred from collective assets into the hands of registered private companies, and the villagers continue to hold shares in the shareholding companies (Zhu 2016; Chung and Unger 2013). This is a bottom-up emergence, which contributes to the jobs-housing co-location process in the study urban village (Fig. 4.1).

4.3.3 Self-Organisation Process Facilitates the Jobs-Housing Co-location Process

A high jobs-housing balance in a neighbourhood will reduce travel time and distance in work trips. The transport literature suggests that the jobs-housing co-location process facilitates the achievement of high jobs-housing balance (Kim 2008; Levinson 1998; Levinson and Kumar 1994). Workplaces and housing mutually co-locate to cut travel times and improve workers' accessibility. To improve their accessibility to work, workers relocate their housing in proximity to employment opportunities. At the same time, service and manufacturing industries change their firm locations in proximity to workers to tap into cheap labour sources. These adjustments might involve a change in residence or a change in the workplace (and sometimes both), which is called the co-location hypothesis (Levinson and Kumar 1994; Gordon and Richardson 1991). The previous section investigated the interaction between rural land acquisition policy and the actions of local villagers in urban villages. The process produces new structural properties, including affordable housing and TVEs and facilitates the jobs-housing co-location process.

Another self-organisation process that facilitates the jobs-housing co-location process in urban villages is the interaction between the *hukou* system and the individual actions of migrant workers. The workers transform the *hukou* system through the reproduction of routine activities to produce new structural properties that include moving swiftly to live near job opportunities in urban villages. The movement of migrants in order to live close to jobs is less costly due to their low level of homeownership and high level of temporary employment status. Migrants can adjust their residences and workplaces more easily to enhance the jobs-housing co-location process (Li and Liu 2016). In 2006, a study on intra-urban migrant mobility in Beijing and Shanghai found that rural migrants' median residential mobility rate was 31.2% per annum (Wu 2006). A study on migrant enclaves in Beijing found that the lower the educational attainment levels the migrants had, the larger spatial distances they experienced in the process of residential and job mobility (Lin et al. 2018). In addition, the household structure of migrant workers influences their residential and job mobility. Migrants who are married and without children or men who are unmarried tend to experience a higher rate of job and residential mobility than migrants who are married with children experience (Lin et al. 2018) (Fig. 4.1).

Other factors that cause the movement of migrants towards employment are the particular attractions of certain neighbourhoods, such as affordable housing and jobs. The building of workplaces in proximity to affordable housing in urban villages is a factor that attracts rural migrants to change their residence in the urban areas and move to urban villages in the inner suburbs. Furthermore, urban policies to suppress secondary industries and develop tertiary ones in the central business districts have encouraged manufacturing workplaces to relocate to ETDZs in the inner suburbs of coastal cities.

For example, a study on the location choice of foreign manufacturing firms in Shanghai found that most Japanese-funded electronic information manufacturing firms are concentrated in the inner suburbs of Shanghai, with only a small number in the outer suburban areas (Jiang et al. 2018). These firms adjust their locations to the inner suburbs because they can not only be located near main transport networks but also tap into the cheap labour sources that are concentrated in the inner suburbs. The change to ETDZ workplaces also facilitates the jobs-housing co-location process (Fig. 4.1). The jobs-housing co-location process produces new structural properties, which include migrants travelling short distances to reach their jobs and migrants using slow transport modes to reach activities. A study on the commuting of migrants in urban villages found that even though urban villages are spatially segregated from formal employment opportunities, migrant workers' average travel time to jobs in urban villages is 4.12 min shorter compared to those of migrants not living in urban villages (Zhu et al. 2017).

4.3.4 Low Educational Attainment Restricts Migrant Women in Finding Jobs

Low educational attainment and labour markets are both types of top-down emergence that contribute to the exclusion of migrant workers from earning adequate income to maintain an adequate standard of living. Educational attainment is a key element of human capital that enables or prevents a person from finding a job. Labour markets follow the rules of society to select and distribute workers among various occupations. The low educational attainment of rural migrants limits their ability to earn an adequate income and enter select occupations. China's traditional culture is based on Confucianism, which stresses the patriarchal role of men. The cultural pressure of son preference in Chinese families determines the dominant status of males in property inheritance, living arrangements, the gendered division of household labour, the continuity of families, and education. In poor rural communities, girls never receive more education than boys do (Jiang et al. 2011). Migrant women always engage in low-paid retail and manufacturing jobs. Girls' education often only reaches the junior secondary school level, and their parents offer marriage as an alternative (Liu et al. 2008) (Fig. 4.2).

In addition, unjust labour markets in urban areas constrain rural migrant workers' choices and freedoms in seeking jobs in urban areas, especially compared to local workers. Most rural migrant women take up low-paying and low-skilled jobs in informal labour markets, and some of them become self-employed. Most migrant workers are unskilled manual or service workers who are paid piece rates or by the hour; thus, many migrant workers put in 19-h shifts in sweatshops (Zhang 2014). As a result, rural migrant women carry out actions, such as working long hours in sweatshops, in order to earn adequate income (Fig. 4.2).

However, the migrant workers' individual decisions and actions have directly influenced their travel patterns to non-work activities and their meaningful participation in society, such as making social contact with local residents and participating in recreation and shopping activities. Sen's theory of exchange entitlement mapping helps to explain the interaction between the unfair labour markets and long working hours of rural migrants in urban China. This is one of the self-organisation processes in urban villages.

Sen's entitlement theory on exchange entitlement mapping (E-mapping) is the concept of the rate of exchange of the entitlement set (e.g., food, housing and income) for the endowment set of an individual (e.g., labour-power and educational attainment) (Sen 1981). The endowment set of rural female migrants includes low educational attainment and labour power, which can only be exchanged for low income in urban China.

The *hukou* system, which is a top-down emergence, negatively influences the E-mapping of migrant workers in urban China. The inequitable labour rules imposed by the *hukou* system, together with the migrants' low educational attainment interacting with the individual actions of rural migrants, produce new structural properties, one of which is the short-distance travel behaviour of rural migrants, who

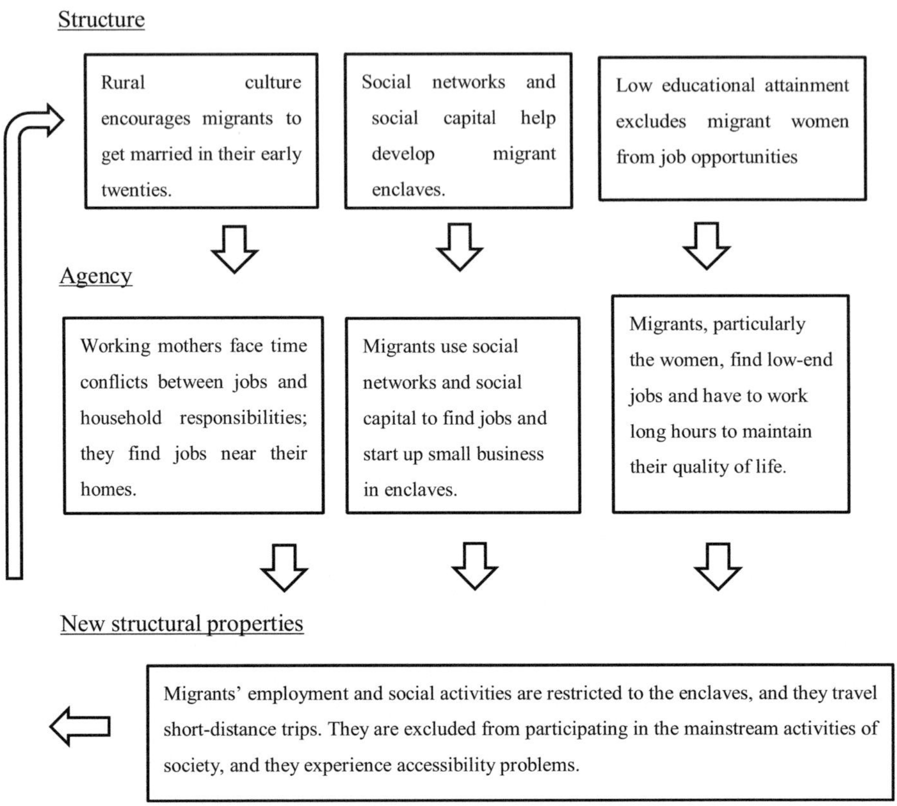

Fig. 4.2 Rural culture, social networks and educational attainment interact with the individual actions of migrants to produce new behaviours

experience time constraints in commuting to non-work-needed activities within a 24-h period of a space-time prism (Hägerstrand 1970). That is, their low educational attainment interacts with their action results in excluding them from participating in non-work activities and accessing job opportunities.

4.3.5 Social Networks Enable Migrants to Find Jobs and Borrow Start-up Cash for Small Businesses

In the social analysis, structures refer to rules and resources, and Giddens defines resources as follows:

> The media whereby transformative capacity is employed as a power in the routine course of social interaction. Giddens (1979, p. 92)

This means that resources are the media that can serve as a source of power in social interactions, such as social networks and social capital, enable social interactions among rural migrants to find jobs or borrow start-up cash for developing their small business in urban villages. The research literature on urban villages indicates that a high percentage of rural migrants find jobs through social networks. The self-organisation process takes place in urban villages. Social networks and social capital are top-down emergence, while the individual actions migrants take to find jobs are bottom-up emergence in urban villages (Li and Zhao 2018; Zhu 2016). The social networks are developed in different formats. Migrants engage more positively in mutual help, and these social networks are based on the origin of the place. They are often concentrated in one urban village to form their own community. Migrants obtain job and housing information through this type of social network (Wu and Logan 2016). The social networks are also used to enable migrants to develop their cultural identity and sense of belonging and even find marriages (Fig. 4.2). Therefore, the social networks interact with the actions of migrants to produce new structural properties, which include their social trips and employment activities that are spatially restricted to the enclaves. As a result, these social networks cause the exclusion of migrants from making social contact with local residents and participating in the mainstream activities of society in host cities.

Another format of social networks is the family support networks or kinship. The choices and actions of rural migrants involve using this type of social network to find jobs and borrow start-up cash for developing small businesses in urban villages. For example, when a rural migrant needs financial capital for an initial investment, he or she usually turns to family members and relatives for help. Two studies, which are based on large-scale surveys on rural-urban migration in China, found that among the interviewed migrant households, 20% were self-employed, and they had mostly started up an array of businesses, from food stalls, hair salons and stationery shops to small manufacturing/processing factories in urban villages. Migrants who run self-employed businesses generally work long hours to increase profits; therefore, they usually do not have adequate time intervals to travel long distances for social and recreation activities, resulting in social exclusion (Zhang and Zhao 2015).

Social capital is inherent in the structure of relations between individuals. It refers to the institutions, relationships, and norms that shape the quality (e.g., trust, helpfulness and friendliness) and quantity of a society's social interactions (e.g., social interactions in an urban village). Social capital enables individuals in society to achieve their intended goals through means such as borrowing start-up cash for developing small retail business. Many skilled and well-connected migrant workers, who have come to urban villages for several years, have increased their social connections in order to start a small business (Fig. 4.2). One study found that rural migrants' entrepreneurship rates (e.g., the number of entrepreneurs as a percentage of total employment) and entrepreneurial entry rates (e.g., the number of new firms that enter a particular industry) surpass those of their urban resident counterparts. This phenomenon indicates that social capital plays an active role in migrant enclaves (Liu et al. 2018) (Fig. 4.2).

Skilled rural migrants develop bottom-up emergence to make use of their skills in the manufacturing and processing industries to start manufacturing workshops. Based on their good social ties, many recruit reliable and cheap labour from their fellow villagers. For example, in an urban village called Dongfeng Village, which is in Guangzhou, most migrants are linked to the garment industry, which feeds the thriving textile markets of Liwan and Haizhu districts. Working residents in Dongfeng Village mostly come from Hubei Province; thus, it is called Little Hubei village. Migrants, who own their small workshops, often recruit their fellow villagers for garment production (Liu et al. 2015).

Therefore, the structures, such as social networks and social capital, interact with migrants' individual decisions and actions, such as actions taken to find jobs, to produce new structural properties that include the migrants' social and employment activities in urban villages.

4.3.6 Rural Culture Encourages Early Marriage for Migrants

In rural communities, the culture, which prescribes that women get married and have children in their early twenties, excludes women from participating in jobs and earning adequate income and constrains their living arrangements in urban China. For many rural migrant women, marriage and motherhood are essential to their gender identity (Roberts et al. 2004; Roberts 2002). Unmarried women are regarded as "old maids", and their fellow villagers socially exclude them.

In 2008 and 2009, a large-scale project called Rural-Urban Migration in China found that among all rural migrants in urban China, 59.9% were men and 40.1% were women. Their average age was 31.1. In addition, over 62.2% of the rural migrants in urban China were married, and approximately 37.8% of them were unmarried (Xie and Chen 2018). Furthermore, a study about rural migrants in Shanghai indicated that among the rural migrant population in the Minhang District, 14.6% of them had moved to the Minhang District from their rural hometowns because they wanted to accompany their spouse. Most of the migrants in this group were women. In the Minhang District, among the migrant women who were between the ages of 15 and 50, approximately 54.5% of them had one child, 31.9% had two children and 4.2% had three children. Thus, most migrant women had to spend time on childcare until their children were grown up (Minhang District Bureau of Statistics 2016). In addition, a large-scale study on rural migrants in Beijing and Shanghai indicated that over 60% of the respondents were married (Wu 2006). Usually, the migrant women had begun their first migration after marriage.

A study conducted by the World Bank (2014) found that the average period of time that a migrant worker stays in a host city is approximately 7–9 years, and only 20% of migrants bring their families, although more than 50% would like to settle in urban areas. In urban villages, the largest single occupational grouping of migrant women is composed of manual labourers working in TVEs in industrial parks. Some of these workers are married women who are not accompanied by their spouse. They live in factory dormitories, where cohabitation with their spouse is unlikely. Certainly, a large share of these factory workers are unmarried women, who are also called '*dagongmei*'. The factory workers must take on the hardest work in manufacturing factories, and most of them have to work 19 h a day to earn adequate income to support their family in their home villages. They always feel exhausted and tired (Zhang 2014). As a result, they travel short-distance work trips. They do not have time intervals for distant non-work activities. They are excluded from making social contact with the locals and accessing distant shopping and recreational activities (Fig. 4.2).

Married women tend to choose retail and service jobs. They live in private rental flats in urban villages so that they can live with their family members (Li and Liang 2016). In two-worker households, the gendered division of household labour always distributes more household duties to women, while migrant men can continue their economic activities with little disruption (Li and Liang 2016; Zhang and Gao 2014). Most of these married women find retail and wholesale jobs or seek manufacturing jobs in small workshops. Some self-employed married women work as street vendors or rent counters at local stores so that they can tackle the time-conflict commuting problems between their household responsibilities and employment (Li and Liang 2016; Zhang 2014). Therefore, the social norms that prescribe that women get married interact with individual actions, such as finding jobs near homes, to produce new structural properties, which include women travelling short distances for work trips. These new travel patterns cause migrant women to be excluded from participating in distant job opportunities, making social contact with local residents, and reaching distant medical and recreation activities (Fig. 4.2).

4.4 The Study Urban Village in Guangzhou: Qingfengcun

The study urban village is called Qingfengcun, which is located in the inner suburbs of Guangzhou and is approximately 17 km from the city centre (Fig. 4.3).

In the Sixth Guangzhou Population Census in 2010, the Shijing sub-district, (which is included the study urban village, Qingfengcun), consisted of a permanent population of approximately 209,022 persons, and approximately 71,337 persons were local *hukou* residents, meaning that the rural migrant population accounted for approximately 65.9% of the sub-district's population (Guangzhou Statistics Bureau 2018). Although Qingfengcun is not near the Guangzhou MTR network, several bus lines link the village with the city centre, whence it takes approximately 45 min to reach the study village.

In 2013, the estimated population of Qingfengcun was approximately 9,000 persons, of whom approximately 4000 were native villagers and 5000 were migrant workers. Qingfengcun is a typical urban village, being located near TVEs and providing affordable housing for migrant workers during the process of the suburbanisation in Guangzhou. The land ownership policy allowed village collectives to reserve industrial land areas for TVE-organised industrial parks. Many large

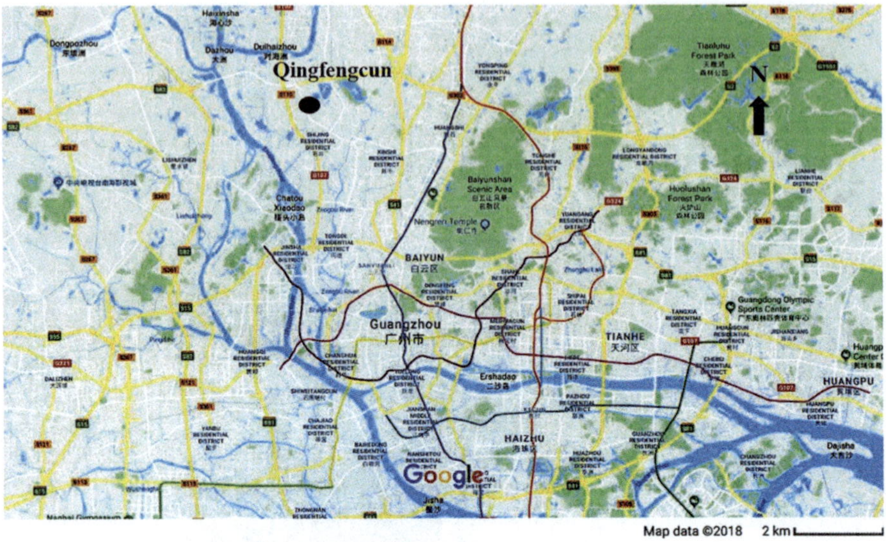

Fig. 4.3 Location of Qingfengcun in Guangzhou. Source: Modified from Google Maps (2018)

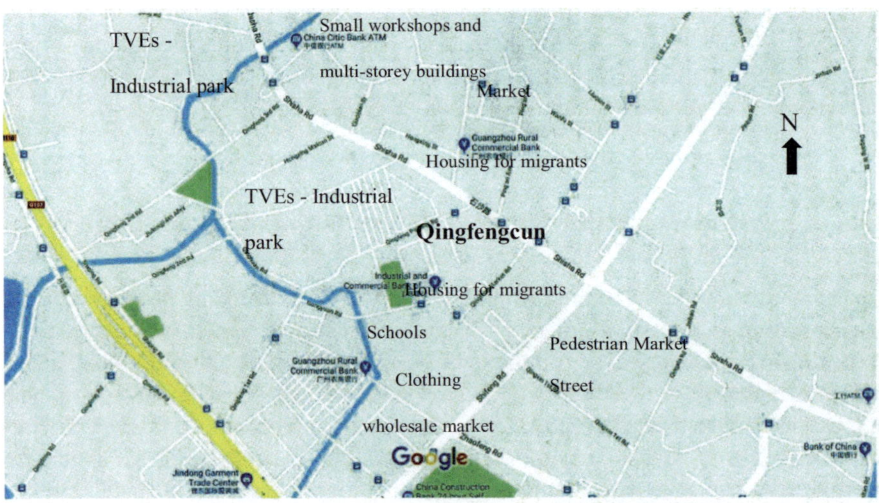

Fig. 4.4 Spatial distribution of the land-use activities in Qingfengcun. Source: Modified from Google Maps (2018)

industrial processing factories had been built in the southern part of Qingfengcun (Fig. 4.4). In addition, some migrants had started small-scale manufacturing and processing workshops in seven-storey buildings, most of which were located along Hongxing Road in the northern part of the village, and these new firms employed approximately 10–20 migrant workers each. With the influx of migrants, Qingfengcun developed into a migrant enclave that provided not only affordable housing but also a range of other neighbourhood facilities such as open markets, a wholesale clothing market, restaurants, schools, convenience shops, barbershops and a public transport interchange. Government public services did not cover the urban village. Village collectives provided informal and private services, such as schools (Fig. 4.4).

In 2014, the author found that Qingfengcun covered an area of approximately 2.35 km^2. Most of the factories in Qingfengcun specialised in footwear production. On the southern and eastern sides of the village, there was also a large-scale wholesale clothing market and a pedestrian shopping street that offered numerous employment opportunities for migrant workers (Fig. 4.4). Because both the factories and the retailing businesses were private enterprises, they could employ migrant workers without regard for the restrictions imposed by the *hukou* system. In addition, because of the collective ownership of land, construction in Qingfengcun was outside urban development control and lacked appropriate site plans. Space was used intensively, and there was a lack of public amenities and green space. As their places of employment were near their homes, most migrant workers commuted by bicycle or walked. The traditional village collectives of Qingfengcun had been transformed into a new form of rural shareholding co-operative that rented out factories to foreign and local investors.

In fact, self-organisation processes take place in Qingfengcun, where social norms, land-use and the *hukou* system interacted with the individual actions of local villagers and migrants to produce new structural properties, which included the formation of jobs-housing co-location processes, short travel distance patterns, the use of slow transport modes to reach activities and close social relationships among migrant residents. Additionally, migrants usually worked for long hours to produce footwear products. They seldom had time to travel long distances to the urban areas, and they did not like developing close social relationships with the locals. The results of the in-depth interviews will further reveal their choices and actions.

4.5 In-Depth Interviews and Travel Surveys

To understand the interaction between the structures and travel practices of the migrant workers in Qingfengcun, in-depth interviews were conducted in 2011. The author conducted the in-depth interviews and engaged in approximately 72 h of observations and conversations with 12 respondents about their actions in interacting with government policies. The results of the in-depth interviews were representative of the types of people living in Qingfengcun, but the data were not statistically representative.

Regarding the questionnaire travel survey, in 2011, a research firm was invited to conduct a survey at public bus stations and open markets and along main streets during rush hour on working days. Two hundred selected working residents in Qingfengcun were successfully interviewed face-to-face. Most of the respondents were rural migrants.

All working migrants in the study urban village were knowledgeable agents. That is, they knew and had a perception of the constraints and consequences of their actions. They made individual decisions and took individual actions to meet their basic needs, such as finding employment and housing (Bohte et al. 2009; Giddens 1981, 1984).

The self-organisation process brought together rural migrants and native villagers to form the migrant enclave, Qingfengcun. Recursive interactions between practices and structures existed in various places, such as industrial parks, housing, small workshops, schools, open markets, and the Qingfeng Pedestrian Shopping Street. To understand the migrants' actions and decisions to transform rules and resources to obtain employment and housing, the interviews had to focus on their attitudes, such as when they expressed that employment was an important factor for them in changing residence and moving to the study urban village (Bohte et al. 2009). Thus, this study investigated the actors' interpretations of their living arrangements, perceived quality of life in the study village and attitudes and decisions to interact with social norms and government policies.

4.5.1 Interview Results: Migrant Workers Tended to Live Near Jobs

The high rate of residential mobility of the migrant respondents was one of the practices that initiated the jobs-housing co-location process to cut travel times. The author interviewed migrant respondents when they were walking around the streets to find jobs and temporary flats to rent in the study village. The respondents expressed that most of the migrants came to Qingfengcun mainly in search of employment opportunities. For example, three young female respondents were interviewed in Hongxing Road, where they were seeking to rent private housing. They said that they came from the rural areas of Hunan Province. They had lived in the inner city neighbourhoods in Guangzhou for 1 year before they came to Qingfengcun. Since they could not afford the high rents in the inner city neighbourhoods and felt job insecurity, they decided to change their residence and move to Qingfengcun to seek affordable housing and stable jobs (Fig. 4.5).

Two migrant women were interviewed while they were searching for jobs in Qingfengcun (Fig. 4.5). They decided to come to the study village because they learned that there were so many job opportunities and it was easy to find affordable housing. Housing affordability is a major problem for them; they wanted to seek jobs in TVEs operated by the manufacturing and processing industries because the factories could provide them with affordable accommodation so they could reduce their housing and transport costs. Before they came to the study village, they had

Fig. 4.5 Migrants find housing and jobs in Qingfengcun. Sources: The author took the photographs

worked in the Haizhu District for 2 years and felt that the rents in the urban areas were unaffordable.

In addition, they had learned that a high percentage of the working migrants in Qingfengcun came from Guangxi Province; therefore, they decided to find jobs in the study village so that they could work with their fellow villagers and obtain social support. Even though the above five respondents were in the most economically active and mobile group (between the ages of 15 and 34), they felt that they had to work long, hard hours so that they could earn adequate income to support their quality of life and send money back to their family members in their home towns. The above respondents' experiences represented migrants' decisions and actions to enact the jobs-housing co-location process in the enclave. They changed residences and sought stable jobs and affordable housing. These were the pushing factors that contributed to making the migrants live in proximity to jobs.

4.5.2 Villagers Built Affordable Housing for Migrants

The Guangzhou municipal government acquired the local villagers' farmland for urban use, and the local villagers were excluded from using their labour and farmland to grow crops in exchange for income and food. In the face of this social exclusion process, they tried to transform the land acquisition policy and find a strategy to earn adequate food and income. The local villagers developed TVEs and built affordable housing for the migrants. Several native villagers were interviewed and described their decisions and actions to develop TVEs and rental housing in Qingfengcun. When they received compensation from the municipal government,

Fig. 4.6 Local villagers built multi-storey buildings along Hongxing road. Sources: The author took the photograph

some respondents opened supermarkets and kindergartens, and some rented out rooms to the migrants in the study urban village. Each native villager was entitled to build a multi-storey building on his own housing plot for rental purposes, and each villager had tried to build as many rentable floor areas on his housing plot as possible to create a source of income. Some rooms on the middle floors of the buildings were converted into workshops for manufacturing and processing industry; some rooms on the ground floor were used for restaurants and stores (Fig. 4.6). The village collectives developed unlicensed clinics on the sides of the main road to help migrants meet their healthcare demands at low costs.

Even though there were no lifts in the multi-story buildings, native villagers always reserved the top floor of the buildings for their families to live because of the high crime rates and unsafe living conditions in the village. The local villagers expressed that it was difficult for them to find jobs in urban areas because they were unskilled workers. In reality, rental income was not evenly distributed among the local villagers in Qingfengcun. A villager said that he did not earn much money from renting rooms and did not have much in savings. He worried about his family's future.

4.5.3 Local Villagers Developed TVEs

Villagers said that the village collectives transformed Qingfengcun into a new form of rural shareholding co-operative. The Shijing Township government and village collectives built and rented out factories and apartments to foreign investors, local villagers and migrants. Some of the TVEs were jointly invested in through shareholding mechanisms. Footwear manufacturing and processing industries accounted for approximately 80% of industrial production in the migrant enclave. The starting funds of the TVEs mainly came from government land acquisition compensation and the district government. The village cadres appointed managers to make key investments and production decisions to support the development of TVEs. Some of the native villagers helped to operate the routine management of the TVEs (Fig. 4.7).

A respondent said that the coordination and mutual adaptation between the local villagers and rural migrants turned Qingfengcun into multiple footwear industrial parks. The migrants started small informal manufacturing and processing workshops, which mainly produced footwear components. These footwear components were sold to the medium-sized factories, most of which were owned by foreign investors who rented the factory buildings from the TVEs. The factories then incorporated the components into their own footwear products. This co-ordination gave rise to industrial clusters of small and medium-sized firms that competed with one another and formed a relatively complete industrial chain to produce footwear products (Fig. 4.7). Many medium-sized TVEs were developed in the industrial park on the southern side of Qingfengcun. In addition, there were many small workshops in the multi-storey buildings run by migrants along Hongxing Road on the north-eastern side of the village (Figs. 4.6 and 4.7). The development of TVEs was the pull factor, providing adequate job opportunities that attracted migrants to live in Qingfengcun and thus initiate the jobs-housing co-location process.

Fig. 4.7 TVEs produced footwear products in Qingfengcun. Sources: The author took the photographs

4.5.4 The Division of Household Labour Distributes More Household Duties to Migrant Women

The interaction between the migrant women's actions and the gendered division of household labour produced new structural properties, including the migrant women travelling short distances in their work trips. These new travel patterns caused the exclusion of migrant women from participating in distant job opportunities and recreational and social activities. This was also a social exclusion process. According to the Sixth Guangzhou Population Census in 2010, the birth rate in the Shijing sub-district (including the study village) was 6.37 per 1000 persons, which included native villagers and rural migrants (Guangzhou Statistics Bureau 2018). The high birth rate was attributed to the high rural migrant population size within the sub-district population. Many two-worker households had young children. The gendered division of household labour allocated more household duties to migrant women than to migrant men. Many married migrant women faced accessibility problems in terms of time-conflicts between childcare and employment. As the author roamed through the streets of the village, he saw many women carrying their young children to different activities, and they were walking at a very quick pace. One respondent said she wanted to send their children to kindergarten classes. Another respondent, who worked in an open market, said that since she had given birth to her daughter, she always carried her child with her to the market.

While she was being interviewed, her child was doing homework at a small desk. The interviewee regarded the village as her home because her family was there. However, she complained that the education policy was unfair to migrants. This was because the local children received free education while she had to pay high education fees for her child, and the fees accounted for a significant share of her household budget. To earn adequate income to support her child's education, she had to work long hours.

Another migrant woman noted that nearly all married migrant women had to take on more household responsibilities than the migrant men did; in fact, this was to the benefit of their families. Thus, she was willing to take on more household duties even though she was a full-time retail worker at a shop along the Qingfeng Pedestrian Shopping Street. She usually woke up early in the morning, helped her child get dressed for school, accompanied her child to the primary school and then walked to her own workplace. She felt very busy every day and did not have time for travelling to recreational activities. She could find almost everything she needed in the enclave. In addition, she choose not to work in the nearby factories because that would require her to spend some of her time for childcare on purchasing groceries and other necessities for her family (Fig. 4.8).

Fig. 4.8 A migrant woman carries her child near a workshop. Sources: The author took the photograph

4.5.5 *Migrants Developed Social Networks in the Study Village*

Social networks were important resources for helping migrants find jobs and housing in the study village. The individual decisions and actions of the migrants in the enclave were undertaken in order to find jobs through their relatives and their fellow villagers. In order to make use of the social networks in the village, most working residents tried to get to know as many of their neighbours as possible so that they could help one another in finding jobs, and these actions increased their sense of local attachment. One woman respondent who had lived in the village for more than 10 years said that women were always active in developing their social networks in the study village; they always helped their neighbours by looking after each other's children. Most residents in the village came from Guangdong, Hunan and Sichuan provinces; they tended to trust in others, do voluntary work, and give informal help to their fellow villagers. The actions of the above respondents in the study enclave are in line with the findings of a study that investigated migrants' self-help behaviour in Beijing (Wu and Logan 2016). Migrants were excluded from making social contact with the local residents; that is, they could not socially integrate into the local society. As a result, they depended more on their local social networks to help their fellows and made a living from informal jobs in urban areas.

Some male migrant respondents felt that they had come to Guangzhou mainly for economic reasons and were not very concerned about social integration. Their social networks were mainly limited to relatives and friends in the study village. Since very few of them could speak Cantonese, the local language of Guangzhou, they were unwilling to participate in community activities. The migrant workers spoke the same languages and shared similar values, and their social activities were spatially restricted to the village. As a result, the migrants travelled short-distance social trips. One of the self-organisation processes that was taking place in the study village was that the interaction between the social networks and individual actions of the migrants produced new travel behaviours that spatially restricted the migrants' social activities to the village, resulting in them taking primarily short-distance trips.

4.6 Questionnaire Survey

The main purpose of the questionnaire survey was to collect data about the travel patterns of migrant workers in Qingfengcun to investigate the self-organisation processes that were taking place in the study village. A total of 200 respondents, most of whom were rural migrants, were successfully interviewed. The structures included the *hukou* system, social networks, dualistic land-use system, gendered division of household labour, low educational attainment, and farmland acquisition policy. The individual actions of the migrants, including working long hours, getting married and bearing children in their early twenties, limiting their social networks to ones within the study village, exhibiting a high rate of residential mobility and finding jobs near their homes in order to resolve time conflicts between jobs and household duties. The processes produced new migrant commuting behaviours that resulted in social exclusion and accessibility problems. Chapter 4 adopts regression statistical methods to measure the accessibility of migrant workers. The questionnaire travel survey aimed to confirm that a survey was the best way to test the self-organisation process in human social systems.

4.6.1 The Jobs-Housing Co-location Process Cut Travel Times

One of the self-organisation processes that was taking place in the village was the interaction between the land-use system, building of affordable housing, development of TVEs and high rate of residential mobility among migrants. The process produced a jobs-housing co-location process to cut travel times on work trips. The new travel behaviours could be reflected in the survey results. The average work commute of all 200 respondents was 15.7 min. In addition, approximately 92% of them lived and worked in Shijing Town, and their average travel time was 14.1 min. The figures indicated that nearly all migrant respondents sought jobs in Shijing Town. The short average travel time of the respondents was the product of the jobs-housing co-location in Qingfengcun. The majority experienced low rates of

Table 4.1 Characteristics of the respondents

Gender	N	Percentage	Travel time (min.)
Men	117	58.50%	15.70
Women	83	41.50%	15.80
Marital status			
Single	65	32.50%	15.54
Married	135	67.50%	15.85
Mode choice			
Walking	106	53.00%	9.5
Riding bicycle	22	11.00%	14.4
Motorcycle	23	11.50%	9.5
Public bus	44	22.00%	35.6
Private car	5	2.50%	9.6
Job referral			
Yes	58	29.00%	16.74
No	142	71.00%	15.35
Age distribution			
18–30	99	49.50%	16.69
31–40	66	33.00%	16.82
41–50	26	13.00%	10.69
51–60	9	4.50%	12.22

homeownership and high rates of temporary work status; they could move swiftly from other districts in Guangzhou to Qingfengcun to find jobs opportunities and produce the co-location process. For example, among all 200 respondents, 56.0% (112 persons) lived in rented apartments, 30.5% (62 persons) lived in self-owned housing and 13.0% (26 persons) lived in their workplaces. Most of the respondents who lived in rented apartments and workplaces were migrants. Those respondents who lived in self-owned housing were mostly local villagers. Furthermore, among those respondents who lived in rented apartments, approximately 36.6% had moved from other districts and provinces to Qingfengcun before they moved into the present apartment. In addition, of those who lived in their workplaces, approximately 53.8% were single and could move easily to take advantage of new job opportunities.

Of all respondents, 49.5% were between the ages of 18 and 30, and their average travel time was 16.7 min. Approximately 33% of the respondents were between the ages of 31 and 40, and their average travel time was 16.8 min. These two groups of respondents were in mobile groups that exhibited higher rates of residential mobility and job changes than other age groups (Table 4.1).

4.6.2 *The Gendered Division of Household Labour Interacted with the Actions of Working Mothers*

One of the self-organisation processes was that the actions of migrant working mothers interacted with the gendered division of household labour to produce new travel behaviours, which included the working mothers commuting short distances and trying to seek jobs that were near their homes in order to cut travel times. Of all respondents, 58.5% were migrant men, and their average travel time was 15.7 min. Approximately 41.5% of the respondents were migrant women, and their average travel time was 15.8 min. One of the top-down emergences that constrained the migrants' commutes was the norm that encouraged migrants to get married in their early twenties. Regarding the marital status of the respondents, 67.5% of them were married, and their average travel time was 15.8 min. Approximately 32.5% of them were single, and their average travel time was 15.5 min (Table 4.1). Of all of the women respondents, approximately 71% had travel times under 16 min. The figures imply that most married migrant women had to take action to tackle the time conflicts and commuting problems between childcare and employment. Approximately 82.5% of the married migrant women chose to work within the enclave, and the figure indicated that they wanted to reduce travel times for their work trips. In addition, approximately 75% of the married male workers had found employment in the enclave (Table 4.1).

4.6.3 *The* Hukou *System and Migrants' Low Educational Attainment Caused Low Income*

The structure, including the *hukou* system and the migrants' low educational attainment, interacted with individual actions of the migrants to produce low income, which also led to the unaffordability of commuting long distances. The migrants lived near their places of employment to save transport costs. The results of the survey indicated that approximately 36.5% of the respondents earned less than 1599 RMB (US$235) per month, meaning that they earned less than 33.4% of the average income of the urban workers in Guangzhou in 2011 (Guangzhou Statistics Bureau 2017). Approximately 34.5% of the respondents earned between 1600 RMB and 2999 RMB (US$441) per month. For those who earned 2999 RMB, their income was approximately 62.1% of the average monthly income of urban workers in Guangzhou in 2011 (Guangzhou Statistics Bureau 2017). The low income of the migrant respondents directly influenced the affordability of transportation costs. To reduce transport costs, they sought jobs near their homes. In addition, the above figures indicate that the *hukou* system interacted with the actions of the migrants and produced new structural properties that caused the exclusion of migrants in accessing formal job markets. Migrants felt difficulties in exchanging their labour

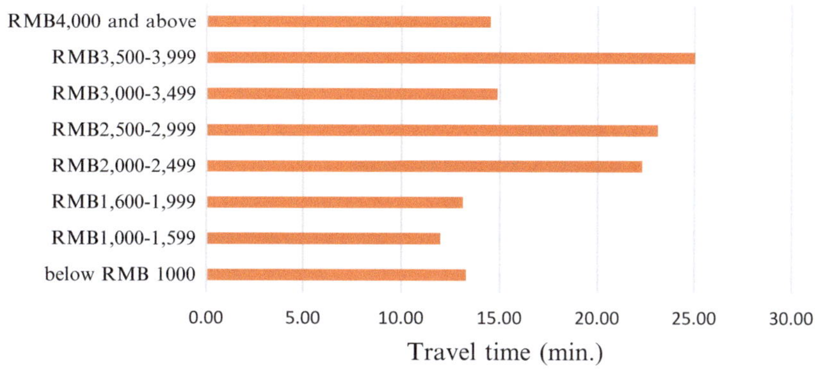

Fig. 4.9 Travel time increased with monthly income

power for an adequate income in order to maintain their standard of living in Guangzhou.

Furthermore, the results of the survey indicated that the average travel times of the respondents increased with their monthly incomes. For example, the average travel time of those respondents who earned less than 1599 RMB (US$236) per month was 11.9 min. The average travel time of the respondents, who earned 2499 RMB (US$367) per month, was 22.3 min. The average travel time of the respondents who earned 3999 RMB per month (US$588) was 25 min (Fig. 4.9).

Influenced by rural culture in China, rural migrant women achieved lower educational attainment than rural migrant men. Thus, on average, the male respondents earned a higher income than the women respondents did. Of all male respondents, approximately 38.5% earned more than 3000 RMB (US$441) or above per month, while only 14.2% of the female respondents earned the same level of monthly income. In addition, approximately 28.2% of the male respondents earned less than 2000 RMB (US$294) per month. Approximately 48.4% of the female respondents earned the same level of monthly income (Figs. 4.9 and 4.10). Thus, on average, the male respondents not only earned a higher income than the female respondents but also had longer travel times to reach their places of employment.

The interaction between low educational attainment and the individual actions of migrants produced short-distance work trips. The low-income respondents found jobs near the enclave in order to cut transport costs, resulting in short-distance work trips. Those respondents wanted to save transport costs so that they could cut their household budgets on transport expenses. Most respondents felt that they needed to work long hours and choose full-time jobs in order to earn an adequate income to maintain their standard of living. Among the respondents, approximately 88.0% sought full-time jobs, and only 12.0% were part-time workers. The average travel times of the full-time and part-time respondents were 15.6 min and 16.5 min, respectively. The above actions imposed commuting constraints on the respondents in accessing distant job opportunities and recreational activities.

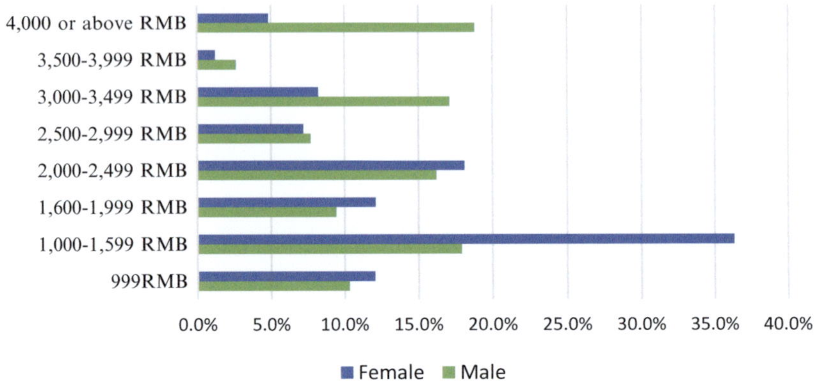

Fig. 4.10 Distribution of monthly income of male and female respondents

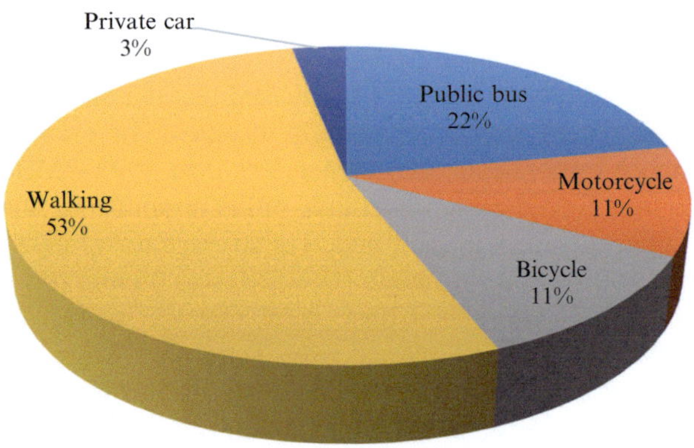

Fig. 4.11 Transport mode choice of migrant respondents

4.6.4 Self-Organisation Processes Produced New Travel Behaviours, Including Migrants Using Slow Transport Modes

The self-organisation processes, e.g., the jobs-housing co-location process, in the study village produced new travel behaviours, which included the respondents using slow transport modes to reach activities. To save transport costs, most of the migrant respondents chose slow and non-mechanical transport modes to reach their places of employment. The results of the survey indicated that of all respondents, approximately 64% used non-mechanical transport modes. Approximately 53.0% of all respondents walked to work, and 11.0% rode a bicycle to work. The average travel times for those who walked or rode a bicycle to work were 9.5 min and 14.4 min, respectively. Of those who took mechanical transport modes, approximately 11% rode motorcycles and 22.0% took public buses (Figs. 4.11 and 4.12). Furthermore,

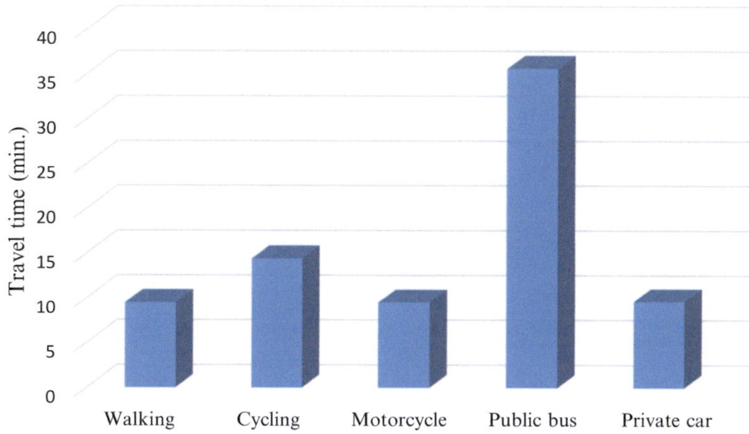

Fig. 4.12 Migrants' travel times and mode choices

among the respondents who took public buses to reach distant employment, 56.8% were migrant men. The figures imply that more migrant men respondents sought distant jobs than did migrant women respondents. In addition, only 3% of all total respondents drove to work.

The transport mode choice figures indicate that most respondents travelled short-distance trips to work, particularly the women. These travel patterns indicated that the self-organisation processes that were taking place in the study village influenced the travel behaviour of the respondents. However, the migrants' behaviour of taking short-distance trips hindered them in commuting to necessary long-distance activities, such as social, recreational and medical activities.

4.6.5 *Self-Organisation Process Produced Behaviour in which Migrants Restrict their Work Trips to the Enclave*

The respondents developed strong social networks and engaged more positively in mutual help, and these social networks were based on the origin of the place. Migrants received job and housing information through this type of social network. The interactions between the social networks and individual actions of the migrants produced the behaviour in which their activity trips were spatially restricted to the study urban village. In accordance with the survey results, of all respondents, approximately 35.2% came from Guangdong Province, 26.5% from Hunan Province, 9.7% from Sichuan Province, and 9.0% from Guangxi Province. Thus, the nature of the enclave made it attractive to new rural migrants from Guangdong, Hunan, and Sichuan provinces. Strong social networks transformed the study village into a migrant-dominated community. It seems that migrants did not depend very much on kinship relationships. Only approximately 29.0% of all respondents obtained their jobs through social networks and referrals, and 71.0% of them sought jobs by applying directly to their employers. Strong social networks enabled the respondents to work and live in the study village, and they spatially restricted their daily trips to the migrant enclave. These travel patterns imposed accessibility constraints on the respondents in accessing distant job opportunities and non-work activities.

4.7 Regression Analysis

Regression analysis was adopted to investigate the impacts of the self-organisation process on the commuting behaviour of the respondents. Variables were chosen for the investigation. An ordinary least squares (OLS) regression model was used to test the strength of the relationships between the independent and dependent variables. The dependent variable of the regression model is the transport mode choice, which is one of the important indicators in representing the travel behaviour of migrants (Table 4.2).

The regression model for mode choice to reach work is as follows:

$$\mathrm{MC} = \beta_0 + \beta_1 * Var1 + \beta_2 * Var2 + \ldots. + \beta 16 * Var16 + \mathrm{E},$$

where MC = mode choice for employment, Var 1 = gender, Var 2 = age, Var 3 = housing tenure, Var 4 = spouse's working location, Var 5 = monthly income, Var 6 = job referral, Var 7 = commuting time, Var 8 = transport fare, Var 9 = job status, E = the error term and β = the regression coefficient.

Figures in Table 4.2 show that among the nine independent variables, four are found to be significantly related to the mode choice variable at the p-value 0.05 level: sex, job status, monthly income, and commuting time. The standardised beta

Table 4.2 Mode choice regression model

		B	Std. Error	Beta	t	sig.
(Constant)		2.313	0.472		4.903	0.000
1	Gender	0.303	0.116	0.296	2.612	0.013*
2	Age	−0.074	0.080	−0.116	−0.927	0.360
3	Housing tenure	−0.020	0.089	−0.027	−0.222	0.826
4	Work location of spouse	0.012	0.012	0.131	1.005	0.321
5	Income	0.109	0.029	0.464	3.794	0.001*
6	Obtain jobs through referrals	0.123	0.122	0.106	1.014	0.317
7	Travel time	0.005	0.002	0.351	3.249	0.002*
8	Travel expense	−0.067	0.041	−0.193	−1.655	0.106
9	Job status	0.337	0.149	0.227	2.261	0.029*

Note: $R^2 = 0.643$, $F = 7.798$, * indicates p-value ≤ 0.05, $N = 200$

coefficients show that job status has the greatest effect on mode choice, followed by sex, monthly income, and commuting time. The other five variables are found to be insignificantly related to the respondents' mode choice to reach work. R^2, which is used to measure the goodness of fit in the regression model, is 0.642, which means that the regression model explains approximately 64% of the variations in the respondents' mode choice to reach work. The sample size for the questionnaire survey was 200 migrant respondents.

4.7.1 The Jobs-Housing Co-location Process Cut Travel Times

Self-organisation processes, such as the jobs-housing co-location process, produced new travel behaviours for the respondents, which included the reduction of travel times to work. Because of the dualistic land-use system, the local villagers could construct affordable, multi-storey buildings on their housing plots and rent these to the migrant respondents. In addition, the village collectives could develop TVEs on their farmland. The migrant respondents had a high rate of residential mobility in order to adjust their place of residence to have access to employment opportunities. This jobs-housing co-location process produced new travel behaviours, which included the migrant respondents taking short-distance trips to reach their jobs. The regression figures show that the travel time variable is significantly related to the mode choice variable. This means that the migrant respondents had short travel times, and approximately 64% of the respondents either walked or rode bicycles to work. The regression figures confirmed that the jobs-housing co-location process occurred in the study village.

4.7.2 Social Networks and Social Capital Interacted with Individual Actions to Produce Short-Distance Trips

The regression figures indicate that the gender variable is significantly related to the mode choice variable. This means that more men walked and rode bicycles to work than did women (Table 4.2). Most of the men and women respondents lived and worked in Shijing Town due to the interaction between their social networks and individual actions. The migrants developed strong social networks in order to find jobs, and social networks were used to enable the migrant respondents to develop their cultural identity and sense of belonging and even find marriage partners. The migrant respondents also developed social capital to help them borrow start-up cash for developing small businesses, such as footwear workshops. Migrants who owned their footwear workshops often recruited their fellow villagers for footwear production. The interaction between the social networks and individual actions of the migrant respondents produced new behaviours in which their routine activities, such as finding jobs and borrowing start-up cash for developing small businesses, were spatially restricted to the study urban village. Therefore, the travel times of the men and women respondents were short.

According to the regression figures, more men used the non-mechanical transport modes to employment than did women. Additionally, the average travel times of the men and women were very close, such as the men's average travel time was 15.7 min and the women's average travel time was 15.8 min. In fact, due to the interaction between their social networks and individual actions, both men and women respondents spatially restricted their work trips to the study urban village. More men could find jobs and start-up small businesses near their homes than could women.

The regression figures also indicated that the impacts of the interaction between the gendered division of household labour and the actions of working mothers on the trip distances of migrant women were overshadowed by the effects produced by the jobs-housing co-location process and the interaction between the social networks and individual actions of the migrants. For example, more men walked and rode bicycles to work than did women, and there was an insignificant difference in average travel times between migrant men and women. However, these findings do not make the effects of the gendered division of household labour any less impactful on the travel behaviours of the women respondents. The interview results indicated that the women chose to take on more household responsibilities than did the men, and the migrant men spent longer amounts of stay time at work than migrant women.

4.7.3 Self-Organisation Processes Influenced the Travel Behaviour of the Full-Time and Part-Time Working Respondents

The self-organisation processes influenced the travel behaviour of full-time and part-time respondents. The regression results show that the job status variable is significantly related to the mode choice variable. The results suggest that a higher percentage of full-time respondents walked and rode bicycles to their jobs than part-time respondents did (Table 4.2). Two self-organisation processes influenced the travel behaviour of part-time and full-time working respondents. The first self-organisation process was the jobs-housing co-location process that cut the travel times of the migrant respondents, who mostly were full-time workers. The local village respondents, who many of whom felt unable to find full-time jobs, were not influenced by the jobs-housing co-location process; thus, they travelled longer distances or had long travel times than the migrant respondents. The local villagers tended to take buses and ride motorcycles to reach employment opportunities.

The second self-organisation process was the interaction between the social networks and individual actions of the migrants. The migrant respondents developed strong social networks in the study urban village so that they could find full-time jobs and earn an adequate income. The local villagers, who did not rely on social networks to find jobs, tended to find part-time jobs in other districts. As a result, the migrant respondents, most of whom were full-time workers, used slow transport modes to reach their jobs, while the local villagers, most of whom were part-time workers, used buses and motorcycles to reach their jobs.

Furthermore, approximately 29.2% of the part-time respondents were local villagers. The results reflect that the local villagers held local *hukou* status, and many of them could choose to seek part-time retail jobs, e.g., in supermarkets and restaurants in urban areas, where they could experience a cohesive working environment with other local workers and earn a higher income. Therefore, most of these respondents took buses to their jobs.

4.7.4 Low Educational Attainment Interacts with Actions to Produce Women's Low-Income

All of the migrant respondents lived in poverty as a result of the *hukou* system. However, the men earned a higher level of income than the women because of the differences in their educational attainment. The low educational attainment of the women constrained their ability to earn an adequate income and afford fast transport modes to reach employment opportunities. China's traditional culture is based on Confucianism, which stresses the patriarchal role of men. The cultural pressure of son preference in Chinese families determines the dominant status male migrant workers in terms of education. In poor rural communities, girls never receive more education than boys. As a result, the women respondents had a lower educational attainment than the men, and the women earned a lower income than the men. As a result, the women could not afford to take fast transport modes to each employment opportunities.

According to the regression figures, the monthly income variable is significantly related to mode choice. On average, the surveyed men earned a higher income than the women. The men also travelled longer distances to reach employment than the women. Thus, some of the men tended to use faster transport modes for long-distance work trips than the women did; for example, they took buses and motorcycles (Figs. 4.9 and 4.10). The regression results confirmed the self-organisation process (such as the interaction between the educational attainment and individual decisions of the migrant respondents), which had produced new travel behaviours, including the male respondents being able to afford to take faster transport modes than the female respondents.

4.7.5 Bottom-up Emergence: To Save Transport Costs

Top-down emergence, such as low educational attainment and the *hukou* system, negatively influenced the ability of the respondents to earn an adequate income. The monthly income of the respondents is significantly related to their mode choice (Table 4.2). One of the bottom-up emergences of the migrant respondents was that they tended to choose slower transport modes to save on transport expenses for work trips. Since there were many retail and manufacturing job opportunities in the study village, many low-income respondents, both men and women, chose jobs near their homes to reduce transport times and costs. Some men, who earned higher incomes than the female respondents, tended to travel longer times to reach employment than the women. The men took faster transport modes to reach distant jobs that could offer them higher wages than they would earn in the study village. The actions of the migrant men are shown in Figs. 4.9 and 4.10.

Furthermore, there were adequate numbers of full-time, low-income manufacturing and retail jobs in the enclave. The migrant respondents, who had low incomes and had to work long hours, tended to seek jobs near their homes so that they could spend more time intervals at work and earn adequate income to support their standard of living. They took slow transport modes (e.g., walking and riding bicycles) to their workplaces.

4.8 Summary of the Survey Results

The results of surveys and interviews indicated that the self-organisation processes produced new behaviours that led to a multi-dimensional phenomenon of social exclusion. The low-income respondents not only were deprived of resources but also experienced a lack of freedom, choice and right to participate in the mainstream activities of society. The *hukou* system excluded the rural migrant respondents from obtaining formal jobs, resulting in the migrants having low income, a lack of education and poor housing. They needed to work long hours so that they could earn an adequate income to maintain their standard of living. As a result, they did not have adequate time to travel to long-distance non–work activities, such as participating in recreational activities and engaging in social contact with the locals. Furthermore, the land acquisition policy interacted with the individual actions of local villagers with the result that the villagers could not exchange their labour power for food and income. Therefore, most of them faced difficulties in participating in the formal job markets and lived in poverty.

The working mothers' actions interacted with the gendered division of household labour, resulting in a lack of capability, freedom and choice to select distant jobs and recreational activities. Migrant women with children did not have adequate time intervals to travel and stay at recreational and social activities. In addition, rural migrant men had higher educational attainment than the women due to the influence

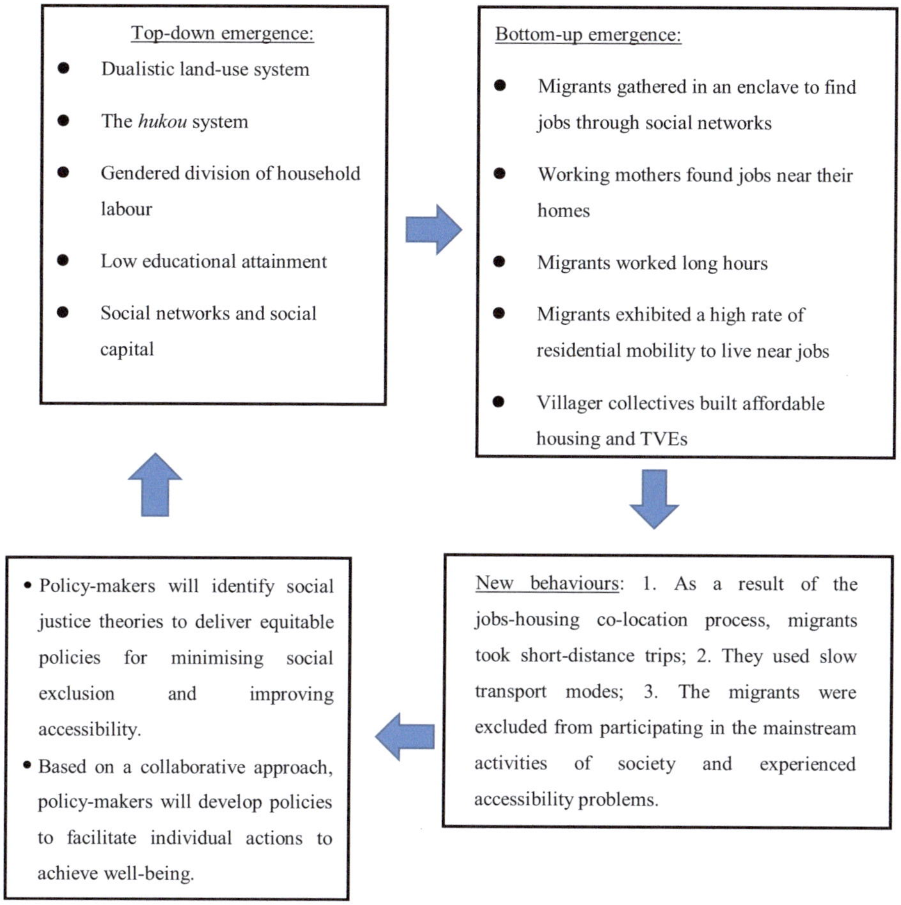

Fig. 4.13 Self-organisation processes produced new behaviours in the study village

of rural culture in China. The migrant women earned lower income and travelled shorter distances to work than the migrant men did. The results of the interviews and surveys confirmed that the self-organisation processes taking place in the study urban village produced a multi-dimensional phenomenon of social exclusion. Most of the migrant respondents were excluded from accessing the mainstream activities of society (Fig. 4.13).

The self-organisation processes produced new travel behaviours in the study urban village, which included the migrants commuting short distances and adopting slow transport modes to reach employment opportunities. Among the self-organisation processes, the jobs-housing co-location process had significant effects on the production of new travel behaviours and social exclusion.

The jobs-housing co-location process involved the interaction between the *hukou* system and individual decisions of migrants to gather together in the enclave, which produced new travel behaviours, such as the migrants choosing to live and work in the study urban village. In addition, the co-location process involved the interaction between the dualistic land-use system and actions of local villagers to produce new structural properties, such as the villagers building affordable housing for the migrants and developing TVEs to tap into the cheap labour sources. The jobs-housing co-location process resulted in the reduction of travel times to places of employment. Migrants also used slow transport modes to reach activities. The co-location process also produced social exclusion problems, such as the migrants choosing to spend a large share of their time intervals on employment and thus not having adequate time intervals for travelling to and participating in distant recreational and social activities.

The interaction between social networks and the individual actions of migrants also imposed significant impact on the commutes and social exclusion of the migrant respondents. The migrant respondents depended on social networks and social capital to find jobs, obtain information on affordable housing and borrow start-up cash in order to develop small retail businesses. The migrant workshop owners recruited their fellow villagers for footwear production in the enclave. Due to their lifestyles, discrimination and language differences, the migrants did not want to make social contact with the local residents. The migrants' routine social networks were limited to their relatives and friends who came from the same hometowns. The self-organisation process produced new structural properties, including migrants' employment and non-work activities, which were spatially restricted to the study urban village. As a result, most migrants did not want to commute long distances to reach social, recreational or medical facilities in other districts (Fig. 4.13).

The impacts of the above self-organisation processes overshadowed the influences of the interaction between the gendered division of household labour and the actions of working mothers in the study enclave. For example, there were no significant differences in travel times between male and female respondents. However, the process did not make the effects of the gendered division of household labour any less impactful on the travel behaviours of the migrant women. The interview results indicated that the women chose to take on more household responsibilities than the men, and the migrant men spent longer stay times at employment than the migrant women did. Most migrant women were married, and they faced time conflicts with commuting problems between household duties and their places of employment. As a result, they did not have adequate stay time to spend on employment, resulting in earning a lower level of income than the migrant men. Furthermore, the migrant women did not have adequate time to travel and to stay at distant social and recreational activities (Fig. 4.13).

4.9 Conclusion

Chapter 4 confirmed the first principal hypothesis of this book: the self-organisation process, which is taking place in Qingfengcun, the study urban village, produces new structural properties as well as accessibility and social exclusion problems. Social exclusion is a kind of self-organisation process and is a multi-dimensional phenomenon that shaped the investigation of the urban village in Guangzhou in this chapter in Guangzhou. One of the dimensions that caused the social exclusion of the rural migrants in Qingfengcun was material deprivation; for example, the migrant workers faced unemployment and earned low income because of the *hukou* system. They had to work long hours to earn adequate income to maintain their quality of life. The migrant workers had to pay high education fees and pay for expensive medical services because they were not entitled to receive public welfare. Furthermore, after the farmland was acquired for urban use, the local villagers could not use their farmland to grow crops, resulting in poverty.

The second dimension that caused the social exclusion of the rural migrants was that the government could not provide them with adequate resources so that they had the ability to achieve well-being. The *hukou* system, low educational attainment and social discrimination caused many migrants to be unable to achieve well-being, resulting in social exclusion.

For example, the rural migrant workers were constrained by the *hukou* system, and they did not have the freedom, choice or opportunity to exchange their labour power for adequate income in formal labour markets in Guangzhou. The Chinese government also followed the utilitarian approach in formulating its policies; such an approach stresses that the best government policy is the one that maximises utility or the benefits to society. For example, the *hukou* system makes use of the cheap labour provided by migrant workers to maximise the GDP of Guangzhou. The system ignores the individual rights of rural migrants to have the freedom to choose their jobs. It also ignores Rawls's difference principle, which states that inequalities are only permitted if they work to the advantage of those who are worst off (Rawls 1971).

The third dimension that caused social exclusion was an unequitable distribution of household duties. Many married migrant women in Qingfengcun were allocated more household duties than the migrant men because of the gendered division of household labour. As a result, the migrant women faced time-conflict commuting problems between their places of employment and their household responsibilities. On average, they earned lower income than the men did.

The other dimension at play is that the migrant workers faced accessibility inequity. Influenced by the *hukou* system, many of them limited their work and social activities to within the migrant enclave, and the migrants were excluded from accessing distant job opportunities and social and recreational facilities, which, together, are the mainstream activities of the society. In addition, local residents' stigma against migrants excluded the latter from participating in certain types of retail and business jobs in the inner suburbs.

In the previous chapters, the author has suggested that structuration theory is the mechanism of the social exclusion process, that is, the interaction between structure and agency (Fuchs 2003; Giddens 1984). The decision-makers of the Guangzhou municipal government can minimise social exclusion and improve the accessibility of rural migrants and local villagers by changing either government policies or individual actions (Fig. 4.13). In order to modify government policies to improve social exclusion, the author suggests the use of social justice theories.

The Rawlsian approach proposes that social and economic inequalities should be arranged so that they are to the greatest benefit of the least advantaged. The approach aims to develop policies to the greatest benefit of the migrants and local villagers so that they can exchange their labour power for adequate food, income and housing in the process of economic reforms in urban China. In addition, based on the capability approach proposed by Sen (1980, 1984, 1992), the municipal government should provide fair policies to ensure that migrants and local villagers have the freedom, choice and opportunity to achieve well-being, including the freedom to choose employment in the labour markets and find affordable housing. Therefore, this book suggests that the Chinese government should transform the *hukou* system so that migrants can find formal jobs and exchange their labour power for adequate income, housing, and medical and educational services.

Under the dualistic land-use system, the farmland in the study village was owned by the village collectives, though they did not have the right to sell the farmland. The farmland in the study village was developed in two ways. First, the Guangzhou municipal government began to acquire farmland for urban development. During the land acquisition process, the government acquired rural land for prices far below market rates. Many local villagers in Qingfengcun became people without land to farm or a job to do and without even minimal social security, and these policies were against the principles of entitlement theory (see He et al. 2009). The entitlement theory stresses that people should have the right to use their land and labour power in exchange for the entitlement set that includes all possible combinations of goods, food and services needed to maintain their quality of life in the open market (Sen 1981, 1984).

The buildings that were constructed by the villagers was of poor quality. Housing was over-crowded, and many roads were not suitable for large public transit vehicles. Therefore, according to the self-organisation process, the municipal government should transform its urban land-use policy in order to help villagers build roads and housing. The government should also provide adequate job opportunities and full social security for the local villagers who were affected by the land acquisition policy so that they can exchange their labour power for adequate income and food.

In addition, top-down policy responses for increasing the social inclusion of migrant workers should be included in government planning solutions and need to be supplemented by bottom-up community participation in order to understand the actions and difficulties of low-income workers and deliver equitable policies for minimising social exclusion and improving accessibility (Fig. 4.13).

A structure is capable of both constraining and enabling action (Giddens 1984). Healey (1997, 1998) proposed a collaborative approach for public participation in order to help policymakers understand the individual actions and attitudes of low-income workers so that the policymakers can modify unjust policies to enable action. The approach includes two types of institutional infrastructure, namely, hard and soft infrastructure. Hard infrastructure includes the consultation of Chinese officials, such as city governments, district governments, street officials, and foreign and local investors, during the process of policy formulation.

Soft infrastructure includes the informal arena: social relations and social networks. Policymakers can consult low-income workers through several informal channels, such as radio and television programmes, and conduct interviews with rural migrants and local villagers about social exclusion issues and their actions to cope with the *hukou* system and land-use policy. Chinese policymakers should tolerate diversity and difference, the heterogeneity of values, and lifestyle oppositions in order to understand the difficulties, skills and capabilities of low-income workers. When policymakers understand the actions and attitudes of the affected workers, they can put the amended policies (e.g., the *hukou* system) into practice.

References

Bohte W, Maat K, Van Wee B (2009) Measuring attitudes in research on residential self-selection and travel behaviour: a review of theories and empirical research. Transp Rev 29(3):325–357

Chan KW (2009) The Chinese *Hukou* system 50. Eurasian Geogr Econ 50(2):197–221

Chung H, Unger J (2013) The Guangdong model of urbanisation, China Perspectives No. 2013/3: 33–41

Fuchs C (2003) Structuration theory and self-organization. Syst Pract Action Res 16(2):133–167

Giddens A (1979) Central problems in social theory: action, structure and contradiction in social analysis. University of California Press, Berkeley

Giddens A (1981) A contemporary critique of historical materialism. Vol. 1. Power, property and the state. Macmillan, London

Giddens A (1984) The constitution of society. Polity Press, Cambridge

Google Map (2018) Google map: Guangzhou 2018. http://maps.google.com/maps. Accessed 2 May 2018

Gordon P, Richardson HW (1991) The commuting paradox. J Am Plan Assoc 57(4):416–421

Guangzhou Statistics Bureau (2017) Guangzhou statistical yearbook 2017, China Statistics Press, Guangzhou (in Chinese). http://210.72.4.52/gzStat1/chaxun/ndsj.jsp. Accessed 9 Sept 2018

Guangzhou Statistics Bureau (2018) The 6th Guangzhou population census in 2010, Guangzhou Municipal Government, Guangzhou (in Chinese). http://www.gzstats.gov.cn/pchb/dlcrkpc/. Accessed 1 Sept 2018

Hägerstrand T (1970) What about people in regional science? Reg Sci Assoc Pap 24(1):7–21

He SJ, Liu YT, Webster C, Wu FL (2009) Property rights redistribution, entitlement failure and the impoverishment of landless farmers in China. Urban Stud 46(9):1925–1949

Healey P (1997) Collaborative planning: shaping places in fragmented societies. UBC Press, Vancouver

Healey P (1998) Collaborative planning in a stakeholder society. Town Plan Rev 69(1):1–21

Hong Kong Trade Development Council (2016) Research: PRD economic profile. http://china-trade-research.hktdc.com/business-news/article/Facts-and-Figures/PRD-Economic-Profile/ff/en/1/1X000000/1X06BW84.htm. Accessed 14 Apr 2018

Huang D, Huang Y, Zhao X, Liu Z (2017) How do differences in land ownership types in China affect land development? A case from Beijing. Sustainability 9(123):1–18

Jiang QB, Li SH, Feldman MW (2011) Demographic consequences of gender discrimination in China: simulation analysis of policy options. Popul Res Policy Rev 30:619–638

Jiang HN, Xu W, Zhang WZ (2018) Transportation accessibility and location choice of Japanese-funded electronic information manufacturing firms in Shanghai, sustainability. Baseline 10 (2):390–412

Kim C (2008) Commuting time stability: a test of a co-location hypothesis. Transp Res A 42:524–544

Levinson DM (1998) Accessibility and the journey to work. J Transp Geogr 6(I):I I–I21

Levinson DM, Kumar D (1994) The rational locator: why travel times have remained stable. J Am Plan Assoc 60(3):319–332

Li C, Liang Z (2016) Gender and job mobility among rural to urban temporary migrants in the Pearl River Delta in China. Urban Stud 53(16):3455–3471

Li SM, Liu Y (2016) The jobs-housing relationship and commuting in Guangzhou, China: *Hukou* and dual structure. J Transp Geogr 54:286–294

Li SX, Zhao PG (2018) Restrained mobility in a high-accessible and migrant-rich area in downtown Beijing. Eur Transp Res Rev 10(4):2–18

Lin WS, Feng J, Li Y (2018) Influence of ICT on housing and employment related migration space of residents in urban villages: a case study of five urban villages in Beijing. Prog Geogr 37 (2):276–286. (in Chinese)

Liu R (2015) Demographic profile, spatial mobility and residence of Beijing's in-migrants: data from the 2010 census. In: Liu R (ed) Spatial mobility of migrant workers in Beijing, China. Springer International Publishing, Cham, pp 125–165

Liu YT, Wu FL, He SJ (2008) The making of the new urban poor in transitional China: market versus institutionally based exclusion. Urban Geogr 29(8):811–834

Liu Y, He S, Wu FL (2010) Urban villages under China's rapid urbanization: unregulated assets and transitional neighbourhoods. Habitat Int 34:135–144

Liu Y, Li ZG, Liu YQ, Chen HS (2015) The growth of rural migrant enclaves in Guangzhou, China: agency, everyday practice and social mobility. Urban Stud 52(16):3086–3105

Liu CY, Ye L, Feng B (2018) Migrant entrepreneurship in China: entrepreneurial transition and firm performance. Small Bus Econ:1–16

Minhang District Bureau of Statistics (2016) Survey on the characteristics of rural migrant workers in Minhang District, Minhang District, Shanghai Municipal Government. http://www.stats-sh.gov.cn/html/fxbg/201605/288460.html. Accessed 7 June 2018 (in Chinese)

Portugali J (2000) Self-organization and the city. Springer, Heidelberg

Rawls J (1971) A theory of justice. Oxford University Press, Oxford

Roberts K (2002) Female labor migrants to Shanghai: temporary "floaters" or potential settlers? Int Migr Rev 36(2):492–519

Roberts K, Connelly R, Xie Z, Zheng Z (2004) Patterns of temporary labour migration of rural women from Anhui and Sichuan. China J 52:49–70

Schwanen T, Lucas K, Akyelken N, Solsona DC, Carrasco JA, Neutens T (2015) Rethinking the links between social exclusion and transport disadvantage through the lens of social capital. Transp Res A 74:123–135

Sen A (1980) Equality of what? Choice, welfare and measurement (Cambridge, MA: Harvard University Press), 1997 [Originally published in The Tanner Lectures on Human Values, Vol. I, by the University of Utah Press and Cambridge University Press, 1980]

Sen A (1981) Poverty and famines: an essay on entitlement and deprivation. Clarendon Press, Oxford

Sen A (1984) Resources, values and development. Oxford, Basil Blackwell

Sen A (1992) Inequality reexamined. Oxford University Press, Oxford

Sen A (2000) Social exclusion: concept, application and scrutiny. Social development papers no 1. Office of Environment and Social Development, Asian Development Bank, Manila

Stanely J, Stanley J, Balbontin C, Hensher D (2019) Social exclusion: the roles of mobility and bridging social capital in regional Australia. Transp Res A Policy Pract 125:223–233

Stanley J, Vella-Brodrick D (2009) The usefulness of social exclusion to inform social policy in transport. Transp Policy 16:90–96
United Nations (2016) Identifying social inclusion and exclusion. In: Leaving no one behind: the imperative of inclusive development, report on the social situation 2016. Department of Economic and Social Affairs, New York, pp 17–28
World Bank (2014) *Urban China:* toward efficient, inclusive, and sustainable urbanization, development research center of the state council, the People's Republic of China. World Bank Group, Washington, DC
Wu WP (2006) Migrant intra-urban residential mobility in urban China. Hous Stud 21(5):745–765
Wu FL, Logan J (2016) Do rural migrants 'float' in urban China? Neighbouring and neighbourhood sentiment in Beijing. Urban Stud 53(14):2973–2990
Xie SH, Chen J (2018) Beyond homeownership: housing conditions, housing support and rural migrant urban settlement intentions in China. Cities 78:76–86
Zhang N (2014) Performing identities: women in rural–urban migration in contemporary China. Geoforum 54:17–27
Zhang CH, Gao Q (2014) The impact of rural-urban migration on gender relations of urban-urban migrant households in China. J Res Gend Stud 4(2):183–205
Zhang Z, Wu X (2017) Occupational segregation and earnings inequality: rural migrants and local workers in urban China. Soc Sci Res 61:57–74
Zhang J, Zhao Z (2015) Social-family network and self-employment: evidence from temporary rural–urban migrants in China. IZA J Labor Dev 4(4):1–21
Zhao P, Howden-Chapman P (2010) Social inequalities in mobility: the impact of the *hukou* system on migrants' job accessibility and commuting costs in Beijing. IDPR 32(3–4):363–384
Zhu P (2016) Residential segregation and employment outcomes of rural migrant workers in China. Urban Stud 53(8):1635–1656
Zhu PY, Zhao SN, Wang LP, Yammahi SA (2017) Residential segregation and commuting patterns of migrant works in China. Transp Res D 52:586–599

Chapter 5
Social Exclusion Produces Commuting Problems and Deprivation in Xiguan, Guangzhou

5.1 Introduction

A number of studies have examined the social exclusion and commuting problems of low-income residents in deprived inner cities, but few have considered how the social exclusion process produces poverty, commuting problems for residents, and the deprivation of capability or, more specifically, the life changes for low-income working mothers and migrant workers. This chapter aims to fill that knowledge gap in urban transport planning with empirical research that can then be used to further understand the transport-related social exclusion problems of low-income workers in urban China. Xiguan, a historic and deprived inner city in Guangzhou, was chosen as the study neighbourhood.

One of the most prominent explanations of the accessibility problems of inner-city poor in the US is the spatial mismatch hypothesis, which posits the exclusion of low-income working residents from accessing labour markets. The hypothesis states that low-skilled and low-income workers are concentrated in the obsolescent inner cities and that most relevant job opportunities are located in suburban areas. The urban decentralisation policy fails to attract suitable job opportunities to locations near low-income communities and leads to job sprawl.

Many low-income workers become unemployed because they do not want to travel long distances from the inner cities to their jobs (Kain 1968). In a jobs-housing spatial mismatch accessibility analysis, transport planners generally focus on the influences of land-use and the main socio-economic characteristics of the population on job accessibility, such as the unaffordability of goods and services for low-income workers, the unavailability of public transit services and poor workers' low-level car-ownership, which hinders long-distance trips (Litman 2017).

Based on the concept of the social exclusion process, the jobs-housing spatial mismatch is due not only to the transport system and built environment problems but also to other structural factors and individual actions. The mismatch is also related to

J. C.-y. Lau, *Self-Organisation Shapes Travel Behaviours and Social Exclusion in Deprived Urban Neighbourhoods of China*,
https://doi.org/10.1007/978-981-15-2252-9_5

other government policies, such as racial segregation policies in housing markets, particularly between blacks and whites; stigma and social discrimination from whites; the fact that black social networks fail to provide information on distant jobs; and non-white workers feeling reluctant to search for jobs in white areas (Ihlanfeldt and Sjoquist 1998).

In addition, individual decisions and actions should play a role in the likelihood that individuals in inner cities have the freedom to obtain jobs. For example, low-income individuals being able to decide to attend school and achieve high educational attainment will help them improve their employability and find suitable jobs (Houston 2005). In addition, policymakers should implement urban redevelopment policies to promote economic opportunities in inner cities. In fact, the spatial mismatch problem is a self-organisation process that produces new structural properties, including a high level of unemployment in inner cities in the US.

This chapter focuses on the investigation of the social exclusion process that is taking place in Xiguan, an inner city in Guangzhou. The commuting problems and social exclusion experienced by the residents in the study inner city are completely different from those of the poor residents in inner cities in the US. The culture, social norms, government policies implemented by the Chinese Communist Party, and individual decisions and actions of people in China are significantly different from the equivalent structural factors in US cities. The rural migrants in Xiguan do not experience spatial mismatch commuting problem because the Guangzhou municipal government is revitalising the economy of the inner city, and the revitalisation policy provides adequate informal jobs for the poor residents in Xiguan.

In addition, low-income residents experience social exclusion problems in Xiguan that are different from those of poor workers in inner cities in the US. The residents in Xiguan live in a deprived inner city neighbourhood with a high number of dilapidated houses that lack essential facilities such as kitchens, flush toilets and sewage disposal systems. The residents live in the inner city not because of racial segregation in the housing market but because of housing and economic reforms and the *hukou* system in China. Unlike poor workers in inner cities in the US, the residents of Xiguan do not need to travel long distances to find jobs because jobs developed near their homes, and most residents take short-distance trips and walk or ride bicycles to reach jobs and non-work activities (Zhou et al. 2015; Yuan et al. 2011). However, the results of the studies in this chapter found that the low-income residents in Xiguan are excluded from participating in the mainstream activities of society. The unjust policies restrict the residents' freedom and opportunity to achieve well-being.

According to the structuration theory, which is the mechanism of the social exclusion process, the structural properties of the social systems in Xiguan are both the medium and the outcome of the low-income workers' practices (Giddens 1981, 1984). Thus, this chapter will investigate the social exclusion processes that are taking place in Xiguan to help urban planners and policymakers understand the constraints of government policies and social norms on the changes in low-income residents' lives. In addition, the chapter investigates the individual actions of residents who reproduce and transform unjust policies and norms to produce new structural properties in order to achieve their intended goals. The new structural properties include the residents subdividing their large houses into small rooms to resolve their housing problems, finding jobs in Xiguan and developing migrant enclaves in their neighbourhood.

The social exclusion process is a self-organisation process that obstructs low-income workers' freedom and right to achieve well-being. Social exclusion is also a multi-dimensional phenomenon. For example, the interactions between the individual decisions and actions of local residents and the economic reforms produce new structural properties, which include the obstruction of the residents' rights to exchange their labour power for food and income; the interactions also lead to the residents' poverty. These are the dimensions of poverty and capability exclusion. The interactions between housing reforms lead the residents to subdivide their houses into small rooms to accommodate more people, which is a dimension deprivation that causes exclusion. The outdated road systems in Xiguan prevent residents from taking public transit and prevent them from moving about freely to seek jobs and reach social and recreational activities. This is the dimension of accessibility inequity that causes exclusion.

The *hukou* system interacts with individual actions of the residents to produce social discrimination against rural migrants and damage the harmonious social relationship between the locals and migrants in Xiguan. The processes also restrict the residents' freedom, opportunity and choice in accessing the social, educational and recreational activities that are available to the majority of people in Guangzhou (Yuan and Wu 2014). Policy-makers can change government policies and use social justice theories to deliver equitable policies to minimise social exclusion. The strategy will also be investigated in the latter part of this chapter. The hypothesis of Chap. 5 is that the social exclusion process that is taking place in Xiguan produces new behaviours, poverty, accessibility inequity and capability deprivation.

In this chapter, the author poses the following three research questions: (1) how does the social exclusion process that is taking place in Xiguan exclude the local *hukou* residents from achieving well-being? (2) How does the social exclusion process that is taking place in Xiguan exclude rural migrants from participating in the mainstream activities of society? (3) Which unjust policies should the Guangzhou government modify to improve accessibility and minimise the social exclusion of low-income workers? Furthermore, it is expected that the investigation of the above three research questions will shed light on research about self-organisation, social exclusion and urban transport planning.

The remainder of this chapter is structured as follows. The next section introduces the study area, Xiguan, which leads to the investigation of the top-down emergence that excludes low-income residents and rural migrants from having the freedom and opportunity to obtain decent homes and access formal job markets. This investigation is followed by a presentation of the results of in-depth interviews with migrants and local residents to explore the influence of the social exclusion process that produces the residents' new behaviours. Then, the results of a survey in Xiguan with rural migrants and local residents are presented. The results mainly focus on how the social exclusion process produces new behaviours that cause social exclusion and accessibility problems. The chapter concludes with a discussion of policies for minimising social exclusion and improving accessibility in this deprived inner-city neighbourhood.

5.2 The Study Area

Xiguan is located near the Zhujiang River and in Liwan district, Guangzhou. The inner city neighbourhood was chosen for study because it not only is an historic inner city neighbourhood in southern China but also has a high concentration of poverty. Starting from the Qing Dynasty (1644–1912), Xiguan (which meant the western gate of the old walled city of Guangzhou) was the main trading port in Southern China for promoting trade and cultural exchange with Western countries. Foreign traders and vendors settled in this neighbourhood. The major trade activities took place along Shisanhan Road, which was located on the eastern side of Xiguan (Fig. 5.1).

Before World War II, Xiguan was still a popular housing neighbourhood for rich businessmen and was famous for its highly established commercial and cultural development, particularly the "Xiguan Big Houses" that were located on Longjin West Road. On the southern side of the area was the Shamian neighbourhood, which once constituted British and French concession in the nineteenth century because of the terms of the Unequal Treaties. The original consulates, churches and banks in

Fig. 5.1 Shisanhan road in Xiguan. Source: The author took the photograph

Fig. 5.2 The Map of Xiguan, Guangzhou. Source: Modified from Google Maps (2018)

Shamian reflected classical Western-style buildings in the nineteenth century. After 1949, the mansions of Shamian became Chinese government offices, hotels or apartment houses.

In any society, people can be considered to be deprived if their standard of living falls below the level attained by the majority of the population. Xiguan is a deprived inner city; studies have found that in terms of education, income, housing and employment, the population in Xiguan experiences high levels of deprivation relative to the other urban districts in Guangzhou. For example, most households do not have their own private toilets (Yuan and Wu 2014; Yuan et al. 2011). This chapter tries to determine the principal causes of deprivation in Xiguan by means of the social exclusion approach. Xiguan covers an area of 6.86 km^2, which includes eight street offices: Changhua, Duobao, Shamian, Lingnan, Haulin, Longjin, Jinhua and Fengyuan. The northern boundary of Xiguan is the Zhongshan 8th Road, and the eastern boundary is Renmin South Road. The southern boundary is the waterfront near the Pearl River (Fig. 5.2).

According to the Sixth Guangzhou Population Census, in 2010, the total population in Xiguan was approximately 300,150 persons, and thus, the population density was approximately 43,754 persons per km^2. The non-local *hukou* population accounted for approximately 34.35% of the total population. The figures indicate that approximately 106,064 rural migrants lived in Xiguan. Among the migrant population, approximately 21,051 persons (19.85%) came from other districts in Guangzhou, and approximately 50,129 persons (47.26%) came from other cities in Guangdong Province, (Guangzhou is the capital of Guangdong Province), and only 34,884 persons (32.89%) came from other provinces in China (Guangzhou Statistics Bureau 2018). In addition, between 30% and 40% of the original local residents had moved out of the neighbourhood. The term working-age population refers to people aged 15–64; there were approximately 233,791 persons in the working-age population, and these persons accounted for 77.89% of the total population in Xiguan. Additionally, there were approximately 38,579 persons aged 65 years or older and approximately 27,780 young children aged 14 years or younger. Thus, the dependency rate was 28.38%. Approximately 48.2% of the total population in Xiguan had junior high school educational attainment or below (Guangzhou Statistics Bureau 2018).

5.3 Social Exclusion Process Represented by Cycle of Agency and Structure

A social exclusion process is a self-organisation process that is shaped by a cycle of structure and agency. The process spontaneously evolves and gives rise to new structures and travel patterns within a social system as a result of the interactions between actors that are not externally controlled, coordinated or regulated (Rauws et al. 2016). The process can be represented by the cycle of structure and agency. The rules or structures shape individual actions, and at the same time, those actions can transform rules through the reproduction of routine practices, producing new structural properties, including social exclusion problems (Fuchs 2003; Giddens 1984). Social exclusion processes that are taking place in the Xiguan neighbourhood include the unintended collective behaviour of local residents interacting with government policies to produce new structural properties.

For example, the planned economy system in place after the establishment of the People's Republic of China in 1949, failed to develop adequate new housing to accommodate the increasing population. The government housing policy (structure) was to transfer the ownership of the old houses in Xiguan from private hands to the state, and these houses were leased to many low-income tenants. In the 1970s, the housing supply in Xiguan did not match the demand following the rapid increase in population. The low-income residents spontaneously transformed the housing policy by turning big houses into small rooms by building partition walls inside each house (agency). As a result, the residents achieved their intended goal of finding affordable housing, but they had to live in over-crowded houses (new structural properties). They were thus excluded from living in a decent living environment (social exclusion).

The structural factors in the social exclusion process include urban redevelopment and revitalisation policies, social discrimination, urban redevelopment policy and housing reforms. These policies interact with the individual actions of residents to produce new structural properties that lead to poverty, accessibility inequity, capability deprivation and spatial segregation between migrants and local residents in Xiguan.

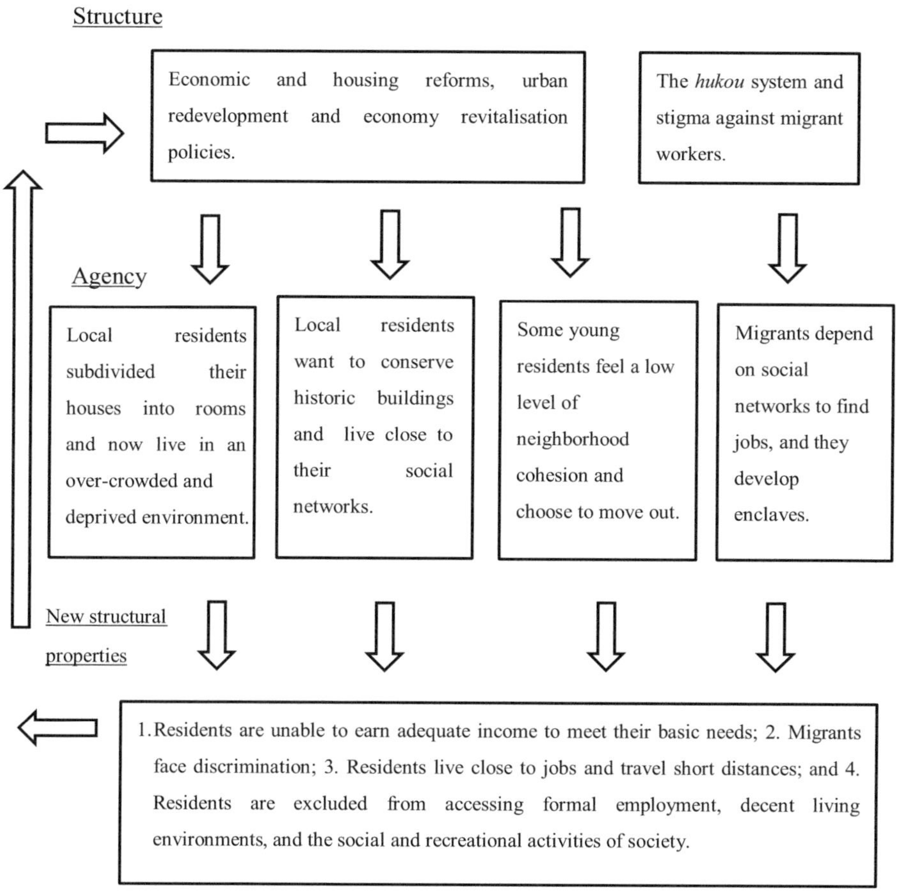

Fig. 5.3 Social exclusion processes in Xiguan

The individual actions and decisions of the poor residents involve changing rules and resources to obtain employment and affordable housing. The process produces new structural properties and travel behaviours. These properties lead to the exclusion of residents from accessing the mainstream activities of Guangzhou. For example, in the face of the high land prices in urban areas and inadequate supply of public housing, local *hukou* residents subdivided their houses into small rooms. The migrants were socially segregated from the local residents, commuted short distances and found accessibility problems in reaching distant social and recreational activities (Zhou et al. 2015). Both the local residents and migrants were unable to exchange their labour power for adequate income and food to maintain their quality of life. The interactions between structures and agencies in Xiguan are depicted in Fig. 5.3.

5.4 Economic and Housing Reforms Cause Deprivation

Economic and housing reforms are the structures that restrict low-income residents from accessing formal employment and decent homes in Xiguan. After coming to power in 1949, the Chinese Communist Party implemented a planned economic system and transformed the private economy into a state-owned or collectively owned economy. Housing in Guangzhou had been mainly provided within a socialist system. Many rich homeowners in Xiguan left China after the Chinese Communist Party took power in China. Other original residents who remained were recruited into collective units to be industrial workers according to the cooperation system between public and private enterprises. In the face of inadequate housing provision, people without housing or workers in work units were allocated vacant houses in Xiguan by the municipal housing bureau. One study found that approximately 60% of the existing residents in Xiguan moved to the neighbourhood in that period of time (Yuan et al. 2011).

One of the top-down emergences were the housing reforms that took place in the 1990s. The reforms abandoned China's old system of linking housing distribution to work units. The subsidised housing traditionally available to state workers was phased out, and work-unit workers had to buy the dwellings in which they lived or turn to the private market to buy commodity housing. Therefore, the municipal government sold publicly owned houses in Xiguan to the sitting tenants at discounted prices during the housing reforms. Many sitting low-income tenants in the inner city became homeowners, but they were also trapped in the deprived neighbourhood. Although the Guangzhou government recognised the importance of housing the poor, the implementation of new welfare measures, such as affordable housing and low-priced rental programmes, constantly faltered due to corruption and a lack of financial resources. Thus, the local *hukou* residents were excluded from accessing decent homes in this period.

Other top-down emergences included the economic reforms that took place after 1978. Most local residents in Xiguan achieved low educational attainment and were low-skilled workers; thus, they mostly worked in collectively owned enterprises. A high number of these enterprises had to shut down because of their poor performance after the economic reforms. In addition, many state-owned factories in urban areas had to be relocated to the suburbs or shut down due to the outward expansion of the metropolitan area. As a result, many residents in Xiguan became redundant workers. Most of them were excluded from accessing formal job markets. In China, approximately 30% (24 million) of all state sector employees were not working in their posts in 1997 (Wang 2000). Gradually, Xiguan became a neighbourhood of concentrated poverty. Furthermore, many migrants moved into Guangzhou to find jobs during that time, and some of them moved to Xiguan. These migrant workers took up most of the poorly paid temporary jobs. The redundant local workers in Xiguan experienced difficulty in finding new jobs. Therefore, they could not afford to buy commodity housing and to move away.

One of the bottom-up emergences for the poor residents in Xiguan was the choice to subdivide their houses and put up with an overcrowded environment because they could not afford to buy commodity housing. For example, the author interviewed a local woman who lived in an old house in Xiguan with her parents. The house was originally allocated to her parents in the 1960s. She once left the neighbourhood to find a job in Yuexiu District. Since she could not afford the high rents, she decided to return Xiguan to live with her parents, and in Xiguan, she found a new job to which she was referred by her relatives. She said that the children of the sitting tenants could inherit the tenant status and continue living in the rental public housing in Xiguan. The old house she lived in was only approximately 45 m^2 (approximately 500 ft^2), but the floor height of the house was approximately 5 m, which allowed her family to build one more apartment in between the floor and the house roof. According to the Sixth Population Census, in 2010, approximately 63% of the households in Xiguan were extended families (Guangzhou Statistics Bureau 2018). Thus, in these households, two to three generations of family members lived under one roof, and their flats were partitioned into tiny bedrooms for their family members (Fig. 5.3). The overcrowding problems often led to social exclusion, such as children lacking the space and peace to do homework and adults lacking adequate space for their social life and privacy.

Furthermore, following economic reforms, the third industry, mainly composed of services, rapidly developed, while the service industry offered a wide range of jobs in urban areas, with wages and salaries varying substantially depending on educational attainment and *hukou* status. Many residents in Xiguan achieved only low educational attainment, and they could only secure jobs in the retail industry that were seasonal, temporary and informal. That is, most local residents were excluded from formal employment opportunities. The income inequality they experienced directly and negatively influenced the affordability of transport modes used to reach employment. For example, Wu et al. (2016) found that a 1% increase in the average wage in urban China is associated with a 1.8% rise in private car ownership. This finding indicates that many low-income workers in Xiguan likely decide to live in proximity to their jobs and to choose the transport modes that match their income. Therefore, some workers ride bicycles to work to save on transport costs (Fig. 5.3).

5.5 The Outdated Road Systems Hinders Commuting

At present, many of the houses in Xiguan have become obsolete, and some building structures are dangerous due to their old age. Lacking redevelopment, Xiguan has numerous dilapidated houses and outdated road systems, some of which were originally designed for horse coaches in the Qing Dynasty. The narrow roads cannot meet the high travel demand for daily work trips. At peak hours, many of the old streets are overcrowded with pedestrians, most of whom are tourists attracted by the commercial streets and a variety of retail markets developed by the urban revitalisation policy. The narrow streets are divided into three portions, in which

Fig. 5.4 Narrow streets in Xiguan are divided into three portions: One each for motor vehicles, bicyclists and pedestrians. Source: The author took the photograph

the bicycle lane is adjacent to the motor vehicle travel lane and flows in the same direction as motor vehicle traffic. Another portion of the street adjacent to the bicycle lane is a sidewalk curb for pedestrians. The streets are usually designed for two-way traffic in order to create higher levels of mobility, but this design also causes traffic conflicts (Fig. 5.4).

Conflicts between motor vehicles and pedestrians regularly happen in the narrow streets during peak hours (Fig. 5.5). The survey results in the latter part of this chapter indicate that some residents, who sought jobs in other districts, have longer average commute times than the middle-class workers commuting in Guangzhou. The outdated road systems, conflicts between pedestrians and traffic, traffic congestion and inadequate provision of transit services contribute to the long commuting times of the residents. Furthermore, this chapter includes a survey on the commuting practices of local residents in Xiguan. The results of the survey indicate that approximately 12.3% of the 171 local respondents surveyed expressed fear of crime in the historic inner city and worries about their personal safety when walking alone in the streets after dark.

It seems likely that many work residents ride bicycles to reach their jobs in this historic inner city because of the high land-use density and more extensive bicycle parking and storage arrangements. In fact, the local residents have been accustomed to riding bicycles to work and non-work activities since the *danwei* era (Zacharias 2012). There are numerous bicycle lanes to connect various nearby activity locations, and the bicycle lanes were originally built for rickshaw and horse carriage

Fig. 5.5 A conflict case between a bus and pedestrians in Xiguan. Source: The author took the photograph

transportation in the Qing Dynasty in 1644. These lanes are for both bicyclists and walkers. The traffic congestion in the neighbourhood causes many work residents to have difficulty fulfilling their basic needs, such as accessing employment and educational activities. Thus, the residents face unmet travel needs problems.

Church et al. (2000) identified six categories that can produce social exclusion: physical impairments, barriers for accessing a given service, affordability, time limitations, fear of crime, and regulatory restrictions. In Xiguan, poor commuting environments, such as narrow roads, dilapidated houses, conflicts between motor vehicle traffic and pedestrians, inadequate public transit provisions, fear of crime after darkness falls and traffic congestion, contribute to the exclusion of residents from commuting to activities that enhance their well-being, such as work, learning, health care, food shopping, and other key activities. Thus, effective and efficient urban redevelopment and economic revitalisation policies are urgently needed in this inner city.

5.6 Social Exclusion and Urban Redevelopment in the Inner City

The urban redevelopment in Xiguan has been a social exclusion process. The process involves top-down planning interacting with the individual decisions and actions of the displaced residents, mass media, and advocacy groups. The process produces new structural properties, including low-income residents continuing to live in over-crowded old houses and other affected residents becoming displaced and lacking access to suburban areas. For example, the Enning Road project in Xiguan, which lasted for the 10 years between 2006 and 2016 (Zhang and Li 2016; Tan and Altrock 2016). The resettlement of the affected residents to the suburban areas resulted in residents lacking access to employment and rail transit services.

There were many dilapidated houses with historic value in the area, and the project had to maintain a low plot ratio to conserve the historic architecture. Thus, private property developers were reluctant to participate in redevelopment projects due to high worker compensation costs and low profits. The municipal government had to take responsibility for most of the redevelopment. The Enning Road redevelopment project started with the Guangzhou municipal government acting on behalf of the state to acquire land and pay an expropriation price to the land users to transform the land into state-owned land, which could then be sold to the property developers for urban redevelopment through open bidding and auctioning (Lin 2015). The municipal government financed the Enning Road Project, and the Liwan District government was responsible for land expropriation, compensation and resettlement.

The owners and sitting tenants of the dilapidated houses in Xiguan did not benefit from the net profits made through the redevelopment of the land. Consequently, there was strong resistance towards redevelopment from existing land users, particularly in regards to the inconvenience of transportation and medical services after reallocation far away from the residents' original neighbourhoods, and endless bargaining for compensation, making it extremely difficult and costly for the urban redevelopment project to break ground (Zhang and Li 2016).

In addition, the original redevelopment plan required the dilapidated houses to be torn down in order to increase profit and the land-use intensity; however, the plans ignored the main objective of preserving the historic architecture and cultural heritage. In the face of tremendous criticism, the government and developers abandoned the original idea of demolition and adopted new plans to conserve the historic buildings and reopen the riverfront (Tan and Altrock 2016). The redevelopment policy attracted higher-income workers to move in and stabilise the decline of the neighbourhood. The reopened river, the historic houses and a museum on Enning Road led to an increase in cultural and leisure facilities. The affected residents were relocated to suburban areas, where they now experience accessibility problems in reaching their places of employment and medical facilities because they live far away from the places needed to fulfil their basic needs.

5.7 Economy Revitalisation Policy Causes Transformation in the *Hukou* Composition of the Population

A number of studies have examined the revitalisation of inner-city historic neighbourhoods in order to build viable businesses that can provide needed nearby employment opportunities for low-income residents. Such revitalisation can lead to the return of high-income groups to the revitalised areas, but few studies have considered how revitalisation shapes the composition of the population in historic neighbourhoods (Porter 2016; McCabe and Ellen 2016; Dokmeci et al. 2007). This section focuses on the influences of the revitalisation of the economy in Xiguan on the population's *hukou* composition and social exclusion.

The social process involves the top-down revitalisation policy implemented by the municipal government and the actions of migrants and local residents in Xiguan. It was a self-organisation process that included the creation of new job opportunities, the inflow of migrant workers and social discrimination against migrant workers (Fig. 5.3). Porter (2016) suggested that economic development in inner cities would come only from recognising the potential advantages of inner cities. These advantages include the historic buildings in Xiguan, the accessibility of the inner-city neighbourhood, the ability to integrate strong regional clusters such as tourism and retail businesses, and an underutilised workforce such as the redundant workers in Xiguan.

Based on Xiguan's convenient centre-city location, with easy access to the Guangzhou Metro networks, and its historic Chinese architecture featuring local Lingnan characteristics, the Guangzhou municipal government instituted a variety of initiatives to revitalise Xiguan. The government promoted retail shopping streets, tourist industries, and a wholesale clothing centre by nurturing private investment. There were already different kinds of retail markets, such as Chinese medicine, goldfish, jade, and clothing markets. These retail markets were developed around the Changshou Road Metro Station and Duobao Sub-districts. In addition, the government renovated the famous large houses along Enning Road and Chinese restaurants in the Changhua Street office to promote the tourist industry (Fig. 5.6).

In addition, the municipal government transformed a group of old houses and streets into the Shangxiajiu Commercial Street, which had an impact on business activities and foot traffic in Xiguan (Fig. 5.5). It was obvious that these businesses had created jobs and had rescued inner-city residents from poverty. Even though the revitalisation policy changed Xiguan from a neglected historic inner city into a vibrant tourist centre that created many retail jobs, the policy failed to attract high-income residents to move into the neighbourhood due to the area's obsolete and dangerous historic buildings and overcrowded living environment.

Living in the inner city after the revitalisation of the economy might provide low-skilled local residents with access to employment through social networks. The neighbourhood effect was that living with other employed adults significantly and substantively increased the residents' likelihood of being employed, indicating the importance of household-based social accessibility for less-educated residents'

Fig. 5.6 The Shangxiajiu commercial street. Source: The author took the photograph

employment outcomes (Parks 2004). Since most of their employed relatives' jobs are restricted to Xiguan, the local residents, who depend on social networks to find jobs, are most likely to find jobs that are located in the inner city, and thus, they take short-distance trips.

In this local-dominated neighbourhood, local residents have had to take action to adjust to the gradual influx of migrant workers. According to the Sixth Population Census, in 2010 a significant share of local residents had moved out of Xiguan (Guangzhou Statistics Bureau 2018). One of the bottom-up emergences of the local residents was renting out their old housing to migrant workers so that they could use the rental income to buy or rent flats in other districts in Guangzhou. Some of them decided to move out of this old and overcrowded living environment. The author interviewed an older local man in Xiguan; he said that he chose to stay in Xiguan because he had a stronger sense of belonging to the community. However, his son had moved out of the neighbourhood because of the overcrowded living environment and dilapidated houses (Fig. 5.7).

Employment instability and low rates of homeownership became push factors to make migrants change their residence and move voluntarily from other urban neighbourhoods to seek employment in Xiguan (Liu et al. 2008). A study in Shanghai found that migrants exhibited a high level of residential mobility in order to find employment. Approximately 49% of migrant respondents reported that they had moved at least once, and approximately 45.7% of the moves had taken

Fig. 5.7 An old man walked along an old lane. Source: The author took the photograph

place in the last 3 years (Shen 2017). Furthermore, migrant-owned businesses always employed more migrant workers than local-owned firms did. This fact suggests that retail business development in Xiguan might increase migrants' employment. In fact, the author found that some types of employment activities, such as the wholesale clothing and retail jobs on the Shangxiajiu Commercial Street, which were characterised by great physical hardship, tediousness, meagre wages, long working hours and low status, were dominated by rural migrant workers (Fig. 5.8). The workers did not have an adequate time budget to spend on commuting to non-work activities, and thus they generated a lower trip frequency than the general working population in Guangzhou. The author also saw several young migrants pushing a large wooden cart to transport heavy clothing products along a narrow road in the Shisanhan clothing wholesale market.

Therefore, the local *hukou* workers have not received many benefits from economic revitalisation. To reduce commuting costs in work trips, migrant workers usually rent out their old houses in Xiguan. Many of them walk or bike to work. Before 1978, labour migration between rural areas and cities in China was prohibited by the *hukou* system. The market reforms in 1978 required a high demand for cheap labour in urban China. The number of migrant workers gradually increased after the "reform and opening-up policy" started. The relaxation of the *hukou* system resulted in migrant workers freely moving to find informal jobs in Xiguan. Thus, the process of revitalisation in Xiguan has contributed to the transformation of the population's *hukou* composition.

Fig. 5.8 A migrant woman transports clothes near Shisanhang road. Source: The author took the photograph

5.8 Stigma Excludes Migrant Workers in Xiguan from Accessing Employment

Stigma is one of the important components in social discrimination; it refers to rules, norms, routines, patterns of attitudes and behaviours in societal structures that constrain the stigmatised groups. For example, 'Mang Liu', (which means blind floating), was widely used in urban China to label migrant workers, and it indicated the stigma against these workers. When the Chinese media first used 'Mang Liu' to describe the large amount of domestic migration initiated in the early 1980s, it implied that migrants were inferior to local residents. The two words literally mean that migrants were headless and did not know where to find jobs and housing to settle down, so they were blindly floating between rural and urban areas (Chen et al. 2011).

One study on rural migrants in urban China stated that stigma is generally considered an attribute of urban residents used to exclude migrants from normalised social activities, including housing and employment. The separation implies a process of devaluation and discrimination against migrants (Li et al. 2006). Moreover, it appears that this social exclusion process always results in mental health symptoms, such as depression, anxiety, hostility and social isolation. The potentially affected population is substantial (approximately one-quarter of the rural labour in mainland China) (Li et al. 2006). In this chapter, stigma is one of the top-down emergences that constrains migrant workers from accessing employment opportunities in Xiguan (Fig. 5.3).

The effects of the stigma against migrants could make it harder for migrants to find jobs in Guangzhou. Their experiences are similar to those of the immigrants in a deprived inner-city neighbourhood in Germany (May 2004). The study by May found that the process of stigmatisation included the legal system in Germany, which assigned the immigrants the position of outsiders. In addition, at the city level, the deprived inner-city neighbourhood of Nordstadt was stigmatised as a neighbourhood for working-class people and immigrants, who were labelled as unclean, disrespectful of norms and prone to violate the law. The immigrants were exposed to stigmatisation and discrimination that led to a general decrease in intergroup trust and social interaction between locals and immigrants.

The immigrants developed individual decisions and actions, such as developing internal cohesion within their respective groups, and made use of the developed social networks to look for jobs and accommodation. In the current study, approximately 38.8% of migrant respondents in Xiguan found employment through their friends and relatives. Some migrants developed strong social capital through which they could borrow start-up cash for developing small businesses in Xiguan (Fig. 5.3).

In fact, the social environment in Xiguan is very similar to that in Nordstadt, Germany. In Xiguan, the local residents compete with the migrants for employment opportunities. Stigmas have developed against the rural migrant workers. The stigmas mainly include two levels of top-down emergence. First, the *hukou* system excludes migrant workers from participating in the formal job markets and social welfare in Guangzhou. The children of migrants have to pay higher school fees in public schools because of their rural *hukou*; they have a strong feeling of a lack of acceptance from their local schoolmates. Another top-down emergence is that stigmas have given rise to a loss of trust between locals and migrants. Migrant workers tend to feel closer to one another if their social distance is small. Social distance is determined by an individual's own socially constructed identity, which is strengthened through shared characteristics with similar values, languages, lifestyles and culture.

Intergroup social trust between migrants and locals was considerably lower than in-group trust in Xiguan. Migrants tended to have low levels of social solidarity, sense of belonging and informal social control (Wang et al. 2017). Migrants and locals generally consider each other out-group members, partly because of the sense of native place identity (lão xiâng). Migrants in Xiguan are often associated with the stigma of criminality, being dirty, having a low level of education, having different lifestyles and languages from locals and being disrespectful of local norms. As a result, social interactions and trust between the locals and migrants has generally decreased. Overall, only a modest degree of social interaction between the locals and migrants appears to exist in the inner-city neighbourhoods. Migrants are excluded

Fig. 5.9 Two migrants selling Xinjiang roast meat on Shangxiajiu commercial street. Source: The author took the photograph

from accessing distant employment opportunities because they find difficulties in developing local social networks to obtain job information.

One of the bottom-up emergences of the rural migrants is that most of them have chosen to find jobs on the eastern side of Xiguan to segregate their own workplaces from those of the local workers. They have developed cohesion within their respective groups in their workplaces to avoid labelling (Fig. 5.9). Most migrant workers choose to live near the Lingnan and Haulin street offices that are located on the eastern side of Xiguan, as they have found retail jobs along the Shangxiajiu Commercial Street and Shisanhang Road (Fig. 5.9). Some rural migrant workers cannot afford the high rents of the houses in Xiguan. They choose to rent affordable flats in nearby urban villages in the Haizhu District and seek jobs in Xiguan. Many of these workers communicate with each other using the languages of their hometowns, and they do not know to speak Cantonese, the local language in Guangzhou.

Most local *hukou* workers choose to work in the Fengyuan, Changhua and Duobao sub-districts on the western side of Xiguan, where they have started up small retail businesses, such as Chinese restaurants, bicycle repair shops, grocery stores, and barbershops. There is a large retail market near the Changshou Road Metro Station. These local workers communicate with one another in Cantonese in their workplaces. Some of them came from cities in Guangdong Province or other districts in Guangzhou, and thus they are considered migrants, but they can speak

Cantonese, the local language in Guangzhou. Many of the original local residents have moved out of the dilapidated neighbourhood, and they rent out their houses to the low-income local workers and migrants, who want to live near their workplaces.

5.9 Questionnaire Survey in Xiguan

A survey was conducted in Xiguan in 2012 to obtain the travel data of the working residents in Xiguan to investigate the impacts of the social exclusion process that took place in the inner city on the commuting patterns of low-income working residents. The population of Xiguan at the time of this study consisted of a high percentage of low-income workers. This dilapidated neighbourhood was surrounded by many new urban developments, and the municipal government seemed unable to redevelop this historic inner city. Instead, many of the streets were left to deteriorate into deprived neighbourhoods. Many of the respondents were migrants, laid-off state workers, or working mothers with low levels of education. The respondents had all experienced social exclusion and accessibility barriers to their travel needs.

Therefore, instead of a conventional transport-planning model, this chapter adopts the concept of the self-organisation approach and social exclusion for investigating the travel behaviour of these social groups. A research firm was invited to conduct interviews near the Guangzhou Metro and public bus stations and in the main streets of Xiguan during rush hour and on working days. The interviews were conducted mainly on Longjin Road, the Zhongshan 8th Road and Duobao Road; all of these roads are located on the western side of Xiguan (Fig. 5.2). Since these roads are in residential neighbourhoods with mostly local residents, a high percentage of the respondents were local working residents. Approximately 220 working respondents were interviewed, and all of them lived in Xiguan. Among the 220 respondents, 170 were local working residents, and 50 were migrants. They were interviewed face-to-face. The respondents were selected randomly within the research neighbourhood.

5.9.1 Survey of Migrant Workers

The purpose of this survey was to investigate the social exclusion process or the recursive interactions between structures, such as stigma and discrimination, and the individual actions of migrant respondents. The process has resulted in new structural properties, including the following: migrant workers take short-distance trips, operate within an enclave that they developed, find jobs through the social network of the enclave, and are excluded from participating in social and recreational activities.

5.9.2 Migrant Residents Take Brief Trips

In the survey, 50 migrants who lived in Xiguan were interviewed, and their average travel time was 25.6 min, which is comparatively shorter than the average travel time of local workers. The shorter average travel time of the migrant respondents to reach their jobs is due to the migrants' actions interacting with the structures, such as migrant women tending to seek jobs near their homes due to the need to take on more household responsibilities. The structures that constrain and enable the commuting of the migrant respondents include the *hukou* system, the gendered division of household labour, social networks and the stigma against the migrants.

5.9.3 Migrant Women Workers Find Jobs Near Their Homes

One of the top-down emergences is the gendered division of household labour that influences the travel times of both migrant men and migrant women. Due to cultural norms, most of the migrant working mothers face time conflicts between their household responsibilities and jobs. The results of the survey indicated that the women choose some form of bottom-up emergence to tackle the time conflicts and commuting problems. They find employment near their homes so that they can cut travel times to reach their jobs and have more time intervals for childcare. For example, of all migrant respondents, approximately 42% were men, and their average travel time was 27.9 min, while approximately 58% of them were women, and their average travel time was 23.9 min. The women reported a shorter average travel time than the men.

5.9.4 Most of the Migrants Have Retail Jobs and Short Time Commutes

Of all migrant respondents, approximately 54% were single, and their average travel time was 24.9 min. Approximately 46% of them were married, and their average travel time was 26.6 min. In addition, among the married migrants, approximately 63.6% had children. The respondents who were single had a shorter average travel time than the married respondents because most single respondents were young workers who mostly sought jobs in food and retail businesses along the commercial walking streets to reduce the travel cost of work trips. Of all respondents, approximately 87.8% were full-time workers, and only 12.2% were part-time workers. Most of the part-time workers were migrant women who wanted to spend more time taking care of their young children and spend fewer time intervals on employment.

Another top-down emergence is the *hukou* system, which excludes the migrant respondents from accessing formal jobs in urban areas. As a result, most migrant respondents have sought informal service jobs in Xiguan. For example, of all of the migrant respondents, nearly 50% had sought jobs in Xiguan, and their average commute time was only 16.6 min, while approximately 36% of them had sought jobs in nearby districts, such as the Haizhu and Tianhe districts.

5.9.5 *Migrant Workers Developed an Enclave on the Eastern Side of Xiguan*

In terms of the distribution of migrant respondents by industry sectors, approximately 26% of the respondents secured informal wholesale and retail jobs in Xiguan, and their average travel time was 15.7 min. Approximately 14.3% of them had found informal jobs in the rental business, and their average travel time was 30 min. Approximately 12.2% of them had secured jobs in the restaurant and food industry, and their average travel time was 27.5 min (Fig. 5.10). The travel time figures indicate that most of the respondents sought employment in informal retail and service jobs along Shangxiajiu Commercial Street and Shisanhan Road, which are located on the eastern side of Xiguan. Furthermore, their travel behaviour indicates that a large share of migrants have sought employment near their homes, where they have developed a migrant enclave on the eastern side of Xiguan.

5.9.6 *Migrants Seek Jobs Through Social Networks*

Because migrants face social discrimination and are excluded from seeking formal jobs by the *hukou* system, they tend to seek jobs through social networks. Of all

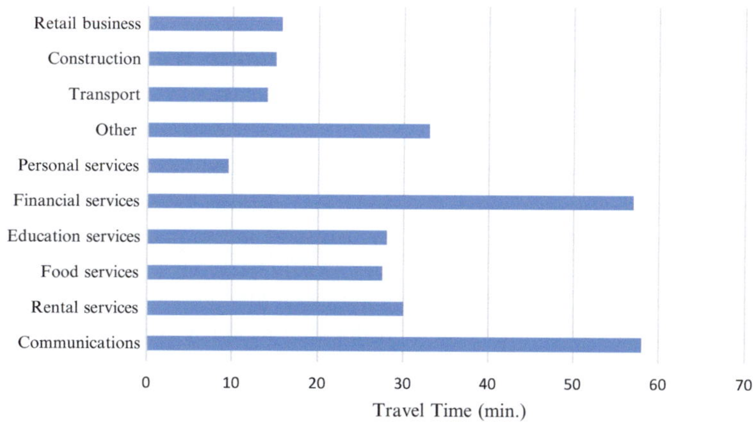

Fig. 5.10 Migrants' travel times by industry sector

respondents, over 64% came from other cities in Guangdong Province (Guangzhou is the capital city), and their average travel time was 24.8 min; the short travel times of these respondents indicate that most of them sought retail jobs in the study area. Approximately 12% of the migrant respondents came from Hunan, and their average travel time was 26.7 min. A significant number of migrant respondents sought jobs that were related to the retail and wholesale clothing industry on Shisanhan Road. Among all migrant respondents, approximately 38% had secured jobs through their social contacts, and their average travel time was 24.7 min, while approximately 62% secured jobs through direct employee recruitment from business firms, and their average travel time was 26.2 min.

5.9.7 Some Migrants Own Their Apartments in Xiguan

Among the migrant respondents, approximately 42% owned their apartments, and this group of respondents had lived in Xiguan for a long period of time, particularly those respondents who came from other cities in Guangdong Province. Thus, this group of respondents exhibited a low rate of residential mobility. They tended to find jobs in Xiguan first and, after a period of time, when they had earned an adequate income, bought their flats in Xiguan so that they could cut the travel cost of work trips. Their average travel was 26.2 min. Of all respondents, approximately 58% rented houses either from private sources or from public housing markets in Xiguan. This group of respondents had lived in Xiguan for a shorter period of time, exhibited a higher rate of residential mobility so that they could live in proximity to jobs, and experienced short average travel times. The average travel time of this group of respondents was 26.3 min. Some migrant respondents had short travel times because they were self-employed and often used their residence as a shop or workshop.

5.9.8 A Large Share of Migrant Respondents Were Young Workers

The previous investigations in this book indicated that the interaction between the *hukou* system and the individual decisions and actions of migrants caused the migrants' high rate of residential mobility. Most of the migrant respondents were young and rented flats; they could move very swiftly from one place to another place to seek jobs. The results of the interviews indicated that among the migrant respondents, the migrants with a high rate of residential mobility were often the mobile group, who were aged between 18 and 30 years. They were always the first group to seek new job opportunities in the retail markets in Xiguan, and thus, they could live near their jobs and reduce their travel times for work trips. Among all migrant respondents, 74% were between the ages of 18 and 30, and their average travel time was 24.5 min. Approximately 12% were between the ages of 31 and 40, and their average travel time was 33.3 min. The results of the survey indicate that young migrant respondents can find jobs and live in Xiguan because they have a higher rate of residential mobility than do migrants aged 31 or older.

5.9.9 Migrant Respondents Used Slow Transport Modes

The social exclusion process produced new behaviours for the migrant respondents. The top-down emergences (such as the *hukou* system, the division of household labour and social networks) interacted with the job-seeking actions of the migrants and produced new structural properties, such as taking short-distance trips and slow transport modes for work trips. Their low income and close social ties made the migrants live and work in the same neighbourhood. Since they lived near their workplaces, they tended to choose non-mechanical transport modes and to take public transit to employment. For example, of all migrant respondents, 33% walked to work, and their average travel time was 11.9 min. Approximately 6% biked to work, and their average travel time was 16.7 min. Approximately 43% took buses to work, and their average travel time was 36.9 min. Approximately 12% took rail transit services, and their average travel time was 29.1 min (Fig. 5.11 and 5.12). The social exclusion process produced new travel behaviours that included the migrants using slow and low-cost transport modes; for example, 39% of them used non-mechanical transport modes, and 54% of them used public transit services.

5.9.10 A Large Share of Migrant Workers Lived in Poverty

Top-down emergences, such as the *hukou* system, low educational attainment and stigma against the migrants, constrain rural migrants' freedom, choice and opportunity to earn an adequate income. Of all migrant respondents, approximately 66% earned no more than 3000 RMB (US$443) per month. This group of migrants earned

Fig. 5.11 Transport mode choices of migrant respondents

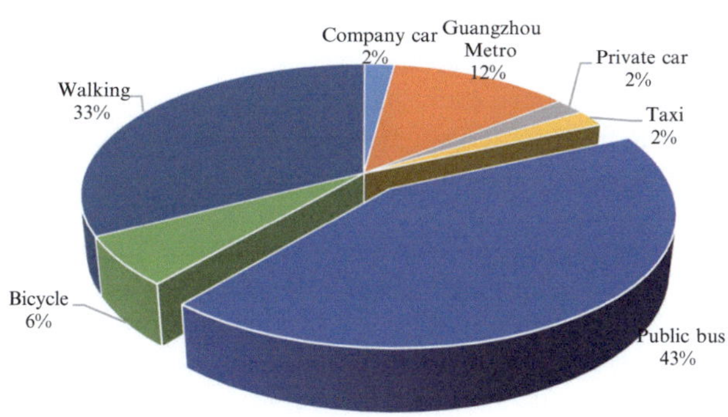

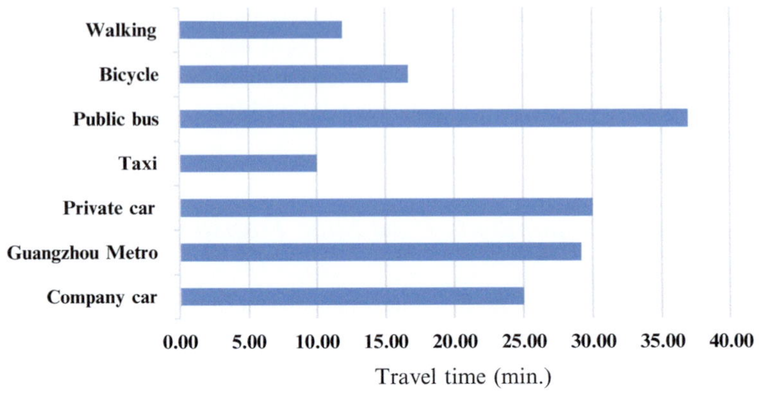

Fig. 5.12 Migrants' average travel time by transport mode

approximately 57.5% of the average city income in 2012, and their average travel time was 24.3 min. Approximately 16% earn more than 5000 RMB (US$740) per month; this group of migrants earned approximately 95.8% of the average income in Guangzhou, and their average travel time was 33.8 min (Guangzhou Statistics Bureau 2017). The above figures indicate that the *hukou* system interacted with the actions of the migrants and produced new structural properties, resulting in the exclusion of migrants from participating in formal job opportunities. The migrants felt difficulty in exchanging their labour power for adequate income to maintain their standard of living in Guangzhou. Additionally, some migrants, who came from other cities in Guangdong could speak Cantonese and were familiar with the lifestyle of Guangzhou, could earn higher incomes than the migrants who came from rural provinces. In addition, the results indicated that the higher the income of the migrant respondent was, the longer his or her commute to work was.

5.10 Survey Results of the Local *Hukou* Workers

The purpose of this section is to investigate the social exclusion process that produced the new behaviours of the local *hukou* residents in Xiguan. This chapter finds that several structural factors have contributed to the multi-dimensional phenomenon of social exclusion that constrains the local low-income *hukou* residents in Xiguan from participating in formal job markets and finding decent homes and suitable living environments. The structures include the economic and housing reform, urban redevelopment and revitalisation policies. The agency or the bottom-up emergences of the local *hukou* residents include subdividing their dilapidated houses into smaller rooms and living in a crowded living environment without privacy. In addition, these residents depend on good social networks to find retail jobs in Xiguan. The interaction between structural factors and individual actions has produced a jobs-housing co-location process in which many of the workers take short-distance trips and walk or bike to work. Because they hold a local *hukou* status, some respondents have been able to find formal jobs in nearby urban districts and thus, on average, they travel longer distances than the migrant respondents.

A high number of local working residents in Xiguan own their houses, and thus, their rate of residential mobility is lower than that of the migrants. The results of these self-organisation actions have produced new structural properties. These new structural properties include the local *hukou* workers living in poverty and being unable to exchange their labour power for adequate food and housing. The local residents have also subdivided their houses into rooms and thus live in over-crowded environments. Many of them can take faster transport modes to work than the migrants can. However, most of them experience congested traffic conditions and conflicts between pedestrians and traffic on narrow roads in Xiguan during peak hours.

5.10.1 A Large Share of Local Hukou *Residents Live in Poverty*

The interaction between the revitalisation policy and the job-seeking actions of the local residents has produced new structural properties, which include the residents finding informal jobs and taking short-distance trips. Thus, the average travel time of all 171 local *hukou* respondents was 29.7 min. This average time was slightly longer than the average travel time of the middle-class residents in Guangzhou. In 2016, the average commuting time of the middle-class residents was 24.73 min (Dai et al. 2016). This was mainly due to serious traffic congestion and the inadequate provision of bus services in the deprived neighbourhood during peak hours. Approximately 73.8% of the local *hukou* respondents spent fewer than 30 min on work trips. In addition, approximately 12.2% of the local *hukou* respondents said that they felt afraid of walking alone at night in Xiguan. Thus, the above figure indicates that the crime rate in the historic inner city contributes to the exclusion of local residents from commuting to activities.

Income inequality exists between male and female local *hukou* respondents. Of all of the female local *hukou* respondents, approximately 35.7% earned less than 2000 RMB (US$296) per month, and of all of the male local *hukou* respondents, approximately 23.9% earned 2000 RMB per month (Fig. 5.13). These groups of respondents earned approximately 37.6% of the average monthly income of all local workers in Guangzhou. In addition, the average travel time of these respondents was 33.3 min. These groups of local *hukou* residents live in absolute poverty. Their income is not adequate for them to fulfil their basic daily needs, such as food and clothing.

Of all of the female local *hukou* respondents, approximately 19.4% earned between 3000 RMB (US$443) and 3999 RMB (US$591) per month, while of all of the male respondents, approximately 26.1% earned between 3000 RMB and 3999

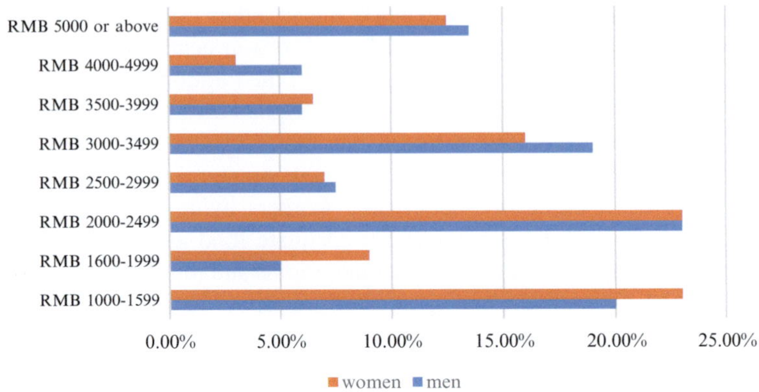

Fig. 5.13 Income inequality between local *Hukou* men and women

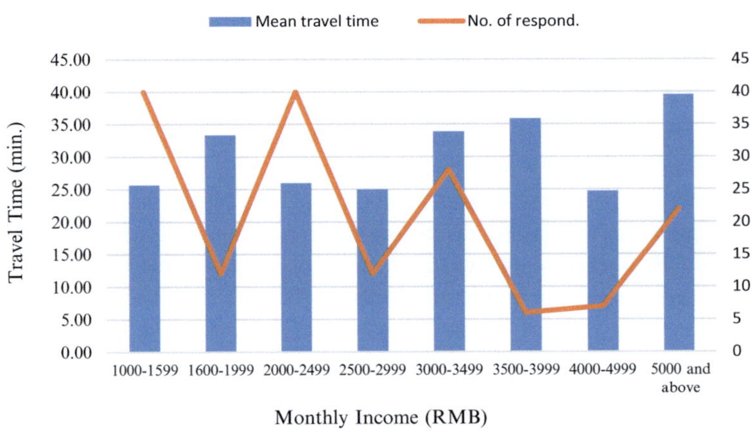

Fig. 5.14 Local residents' travel time by monthly income

RMB per month (Fig. 5.13). These groups of respondents earned between 56% and 75.2% of the average income of all local workers in Guangzhou. These groups of local *hukou* respondents were excluded from earning adequate income due to the impacts of the unjust economic reforms. The average travel time of these two groups was approximately 25 min. The above figures indicate that the social exclusion process or the interactions between the policies of economic reforms and the actions of the local residents in Xiguan have produced new structural properties that have excluded the local residents from accessing adequate income and employment to meet the prevailing standard of living in Guangzhou.

Furthermore, of all female respondents, approximately 12.4% earned 5000 RMB (US$740) or more per month; of all male respondents, approximately 13.7% earned 5000 RMB (US$740) or more per month (Figs. 5.13 and 5.14). The average travel time of these groups was 39.5 min. The local *hukou* respondents with lower income tended to have shorter average travel times than the higher income groups (Figs. 5.13 and 5.14). The government adopted the revitalisation policy to create retail jobs for local *hukou* residents in Xiguan. As a result, many of the informal jobs that are suitable for the local *hukou* respondents are located in the Liwan District, where the tourist walking streets offer many job opportunities. In addition, the above figures indicate that, comparatively speaking, the local *hukou* men earn a higher income than their female counterparts. Therefore, the former travel longer distances and times for work trips than the latter. Furthermore, the respondents who earn higher incomes tend to travel to other districts in Guangzhou; thus, they have longer commutes to work than do the lower-income groups.

5.10.2 Housing Tenure of Respondents Constrained in Moving to Access Jobs

One of the top-down emergences is the housing reforms. During the housing reforms in 1998, the Guangzhou municipal government sold publicly owned houses in Xiguan to the sitting tenants at discounted prices. As a result, many sitting low-income tenants in the inner city became homeowners. In this survey, approximately 68.6% of the local *hukou* respondents owned their flats, and their average travel time was 30.6 min. Approximately 3.5% of them lived in public housing, and their average travel time was 25.8 min. Approximately 4.1% of them rented private housing, and their average travel time was 20 min. The above figures indicate that local *hukou* respondents who own their flats tend to have longer travel times than other groups of respondents. They have lived in the neighbourhood longer than the other groups and thus can obtain distant job information through their established social networks. Additionally, their low rate of residential mobility causes them to have longer travel times between their housing and jobs than the migrant workers in Xiguan.

5.10.3 The Local Hukou Residents Seek Jobs in Formal Job Markets

The local *hukou* residents are not constrained by the *hukou* system; they can find jobs in other urban districts, and this has led to the local respondents having longer travel times than the migrants. For example, among all local *hukou* respondents, approximately 16.3% of the respondents had sought employment in the Tianhe District, and their average travel time was 34.8 min. Approximately 3.5% of them had sought jobs in the Panyu District, and their average travel time was 38.3 min. Approximately 11.6% of them had found jobs in the Haizhu District, and their average travel time was 35.5 min. Approximately 47% of the local *hukou* respondents had sought jobs in the Liwan District, where Xiguan is located. The local *hukou* respondents enjoy the freedom and choice to seek formal jobs in Guangzhou; thus, the *hukou* system has less of an impact on the local residents than on the migrants as far as social exclusion is concerned.

5.10.4 A Large Share of Local Respondents Choose to Find Jobs in Xiguan

The distribution of the local respondents' main industry sectors indicated that most unskilled local *hukou* respondents tended to find retail and service jobs in Xiguan. The survey results indicate that approximately 15.7% of them had sought wholesale and retail jobs in the crowded and touristy streets in Xiguan, and their average travel time was only 18.8 min. Approximately 15.1% of the total respondents had obtained renting and business service jobs, and their average travel time was 25.8 min. Approximately 7% of them had secured food industry jobs in restaurants, and their average travel time was 27 min. Approximately 14.5% of them had found social and personal service jobs, most of which were located in other urban districts in Guangzhou, and their average travel time was 30.2 min (Figs. 5.15 and 5.16). The respondents whose travel times were less than 25 min usually walked or rode bicycles to work, and they most likely had employment within the inner city neighbourhood (Figs. 5.15 and 5.16).

5.10.5 Local Respondents Chose Low-Cost Transport Modes

In terms of the mode choice of the local *hukou* respondents, approximately 4% rode bicycles to work, and their average travel time was 19.2 min; approximately 19% of

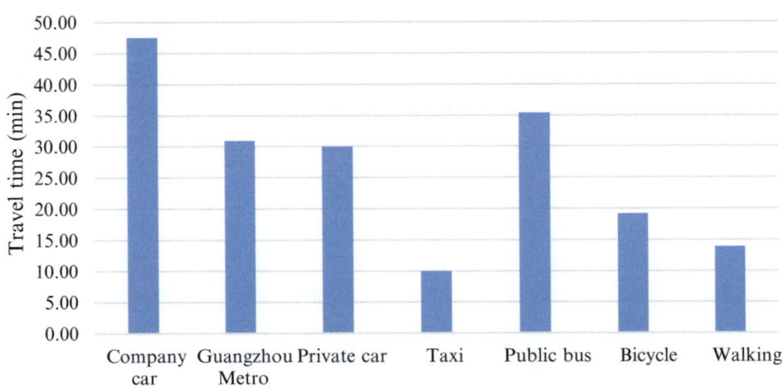

Fig. 5.15 Travel time and mode choice of the local workers

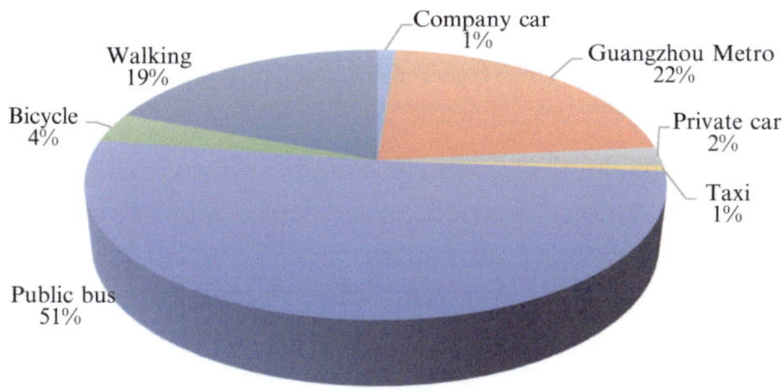

Fig. 5.16 Mode choice of the local workers in Xiguan

the respondents walked to work, and their average travel time was 13.9 min. Approximately 22% of the respondents took rail transit services to work, and their average travel time was 30.9 min; 51% took transit buses, and their average travel time was 35.4 min (Figs. 5.15 and 5.16). According to the results, 74% of the respondents chose low-cost transport modes to reach work, such as walking, riding bicycles or taking transit buses. Transit bus fares for most of the trips in the urban areas were only 2 RMB (US$0.3); thus, most of the low-income respondents chose transit buses for work trips. The respondents who took buses experienced longer trip times than those who took rail transit service because heavy traffic congestion is consistent in Xiguan. In addition, a high number of local *hukou* respondents took public transit to employment, and the new behaviours that were produced by the social exclusion process are different than the new behaviours of the rural migrant respondents. The local residents took faster transport modes and had longer commute times to work than the migrants (Figs. 5.15 and 5.16).

5.10.6 Significant Number of Local Hukou *Respondents Found Jobs Through Social Networks*

Of all local *hukou* respondents, approximately 34.9% found jobs through their social networks, and their average travel time was 27.8 min. Approximately 64.5% of the respondents found jobs through direct employment, and their travel time was 30.7 min. Those respondents who obtained jobs through their social networks had a shorter average commute time than the respondents who found jobs through direct employment. Thus, social networks, or a top-down emergence, enabled the respondents to seek jobs near their homes and cut their travel times to work.

Approximately 56.2% of the local *hukou* respondents were single, and their average travel time was 28.6 min, while approximately 43.8% of the local *hukou* respondents were married, and their average travel time was 30.7 min. The unmarried respondents tended to be younger, and they did not have established social networks to use to obtain distant job information. Thus, they tended to seek retail jobs in Xiguan and had shorter average commute times than the married respondents.

5.10.7 Married Local **Hukou** *Respondents with Children Travel Short-Distance Trips*

Of all of the local *hukou* respondents, approximately 75 respondents (or 43.8%) were married. Among married respondents, approximately 41.4% had young children, and their average travel time was 28.7 min. Approximately 58.6% of the married respondents had no young children, and their average travel time was 31.2 min. Thus, those married respondents without children tended to have longer average travel times that the respondents with young children. These results are in line with the results of the previous studies mentioned in this book. The married workers with children faced time conflicts between childcare and employment, and their bottom-up emergence was to find jobs near their homes and cut their travel times to work in order to save more time intervals for household responsibilities. Some of the married workers sought part-time jobs to resolve the time conflicts and commuting problems. Of all of the local *hukou* respondents, approximately 75% were full-time workers, and 25% were part-time workers. The average travel times of full-time workers and part-time workers were 29.6 min and 29.9 min, respectively.

5.10.8 A Large Share of Local **Hukou** *Respondents Were Young Workers*

Among all of the local respondents, 63% were between the ages of 18 and 30, and their average travel time was 27.8 min. The young age groups tended to seek retail jobs in Xiguan through their social networks. According to the survey results, among the respondents between the ages of 18 and 30, approximately 37.6% had sought jobs through their social networks. Approximately 13.5% were between the ages of 31 and 40, and their average travel time was 38 min. Approximately 14.6% of them were between the ages of 41 and 50, and their average travel time was 30.2 min. Finally, approximately 7.6% of them were between the ages of 51 and 60, and their average travel time was 26.4 min. The travel time figures indicate that the younger local respondents, who were between the ages of 18 and 30, had not established social networks to obtain distant job information but tended to seek retail jobs in the tourist walking streets in Xiguan. They took trips of shorter distances than other age groups.

5.11 Regression Analysis

Regression analysis was adopted to investigate the social exclusion processes that produced new travel behaviours and excluded the local *hukou* respondents from achieving well-being. Ten independent variables were chosen for the investigation.

Table 5.1 Mode choice regression model

	B	Std. error	Beta	t	Sig.
Constant	6.722	1.139		5.903	000
Travel time	−0.012	0.006	−0.143	−1.943	0.054**
Job status	−0.544	0.322	−0.138	−1.689	0.093*
Industry sector	0.016	0.029	0.04	0.554	0.58
Housing tenure	0.372	0.133	0.208	2.798	0.006**
Workplace location	−0.179	0.065	−0.204	−2.736	0.007**
Job referral	−0.528	0.276	−0.147	−1.914	0.057**
Monthly income	−0.127	0.056	−0.18	−2.268	0.025**
Marital status	−0.198	0.351	−0.057	−0.565	0.573
Age	0.243	0.164	0.145	1.486	0.139
Gender	−0.098	0.257	−0.028	−0.381	0.704

Note: $R^2 = 0.227$, $F = 4.574$, ** indicates p-value ≤ 0.05, * p-value ≤ 0.1, $N = 171$; Dependent variable is mode choice: Company coach = 1, the Guangzhou Metro = 2, Private car = 3, Taxi = 4, Public bus = 5, Bicycle = 6, Walking = 7

An ordinary least squares (OLS) regression model was used to test the strength of the relationships between independent and dependent variables. The dependent variable of the regression model was mode choice, which was one of the key indicators representing the new travel behaviours of the local *hukou* residents in Xiguan.

Among the ten independent variables, five are found to be significantly related to the mode choice variable at the p-value 0.05 level, and one is found to be significantly related to the mode choice variable at the p-value 0.1 level. These variables include travel time, job status, housing tenure, workplace location, job referral, and monthly income. Other variables, e.g., industry sector, age, gender and marital status, were found to be insignificantly related to the respondents' choice of transport mode to reach work. The R^2, which is used to measure the goodness of fit in the regression model, is 0.227, which means that the regression model explains approximately 22.7% of the variations in the respondents' choice of transport mode to reach work. The sample size for the questionnaire survey was 171 local *hukou* respondents (Table 5.1).

5.11.1 Income Inequality and Mode Choice

The results of the regression analysis indicated that local respondents with higher income tended to use faster transport modes to reach their workplaces. For example, most of the higher income groups tended to drive private cars, take company coaches or use the Guangzhou Metro. The lower-income respondents tended to walk, ride bicycles or take public buses (Table 5.1). In interacting with the structures, such as the economic and housing reforms, the local respondents tended to choose the action of taking low-cost transport modes, and they chose employment opportunities that were near their homes. Hence, the social exclusion process produced new structural properties, which included seeking jobs near their homes and taking low-cost transit services. As a result, approximately 23% of the respondents used non-mechanical transport modes, and 51% of them took transit buses; a bus fare of US$0.3 allowed the respondents to reach most locations in the urban area. In addition, the local respondents tended to use faster and higher-cost transport modes (e.g., rail transit services) to travel longer distances to reach employment opportunities (Table 5.1). Most of the local respondents were low-income workers, who tended to choose low-cost transport modes in order to save money on transport fares for work trips. As far as social exclusion is concerned, the low-income local respondents were excluded from seeking distant employment opportunities in formal job markets in Guangzhou.

5.11.2 Travel Time and Workplace Locations of the Respondents and their Transport Mode Choices

The results of the regression analysis indicate that those respondents who adopted faster transport modes tended to have longer travel times for their work trips. For example, most of the respondents in higher-income groups tended to drive private cars, use company coaches or take the Guangzhou Metro to their workplaces. Hence, they experienced longer travel times than those respondents in lower-income groups, who tended to walk, ride bicycles or take public buses to work (Table 5.1). In addition, the results of regression analysis found that local residents in Xiguan found jobs in other districts, such as the Yuexiu and Tianhe districts, and tended to drive private cars or take company coaches or the Guangzhou Metro to work. Those who found jobs in the Liwan District or in Xiguan tended to walk, ride bicycles or take public buses to work (Table 5.1).

The above regression analysis results reflect that the respondents who adopted faster transport modes tended to earn a higher income and commute at a faster speed to work than the low-income local workers. The low-income workers experienced accessibility problems in reaching distant job opportunities because they could not afford the costs of fast transport modes. Hence, the high-income residents adopted faster transport modes and travelled longer distances to other districts in Guangzhou to seek higher-paying jobs than the low-income residents.

The structural factors (such as economic reforms, social networks, educational attainment and gendered division of household labour) interacted with the individual actions of the local residents and produced new structural properties. These new properties included the low-income working women earning a lower income than their male counterparts, and the women tending to commute shorter distances to work. The young working residents could not establish strong social networks to obtain distant job information; they tended to seek jobs near their homes and take slow transport modes to work. These groups of local residents did not have the freedom or choice to seek distant employment and recreational activities. They were excluded from participating in the mainstream activities of society.

5.11.3 Job Status and Transport Mode Choice

The regression analysis results indicate that the local respondents who walked and took buses to work were mostly full-time workers, while the respondents who sought part-time jobs tended to take public transit services, such the Guangzhou Metro (Table 5.1). In the social exclusion process, the bottom-up emergence of the married local respondents was finding part-time jobs so that they could tackle the time conflict problems between household duties and employment. As far as possible, most of the respondents tried to find part-time jobs in the neighbourhood in which they lived. The average travel times of the full-time and part-time respondents were very close, e.g., the full-time workers' average travel time was 29.6 min, and the part-time respondents' average travel time was 29.9 min.

It is possible that the interaction between the government revitalisation policy and individual actions of the respondents produced new structural properties, which included the creation of many full-time retail jobs in Xiguan. Thus, many respondents could seek full-time jobs through their social contacts. Most respondents with low educational attainment and low skills sought full-time retail jobs that were suitable for them. Most part-time job opportunities were located in Xiguan, but some were located in nearby districts. Therefore, the bottom-up emergence of married local respondents who wanted to seek part-time retail jobs was taking the Guangzhou Metro and public buses to other districts to secure part-time jobs. The structural factors, such as low educational attainment and the gendered division of household labour, contributed to the exclusion of the part-time respondents from seeking full-time jobs in Xiguan.

5.11.4 Characteristics of the Respondents' Housing Tenure and Their Transport Mode Choice

Housing tenure refers to the financial arrangements under which someone has the right to live in a house or apartment. The regression analysis results indicated that those local respondents who owned their flats or lived in collectively owned housing tended to take the Guangzhou Metro or drive to work, while those who rented public housing or rented other people's private housing in Xiguan tended to walk, ride bicycles or take public buses to work (Table 5.1). The above figures indicate that those local *hukou* respondents who owned their flats tended to have longer travel times than the low-income respondents.

The interactions between the structure and individual actions produced new structural properties, which include the respondents who owned their houses tending to have and use strong social networks to obtain distant job information and seek distant employment opportunities. The local respondents who owned their houses were constrained by the high cost of moving to live near job opportunities and of cutting travel times. Those respondents who rented flats had a higher rate of residential mobility because they did not need to pay as high a cost to move to live near job opportunities and thus they could walk or bike to work. They also wanted to seek jobs near their homes to reduce transport costs. The low income of the respondents contributed to the respondents' exclusion from accessing distant job opportunities.

5.11.5 Finding Jobs Through Social Networks and Transport Mode Choice

The regression analysis results indicate that those respondents who sought jobs through direct employment tended to take faster transport modes, e.g., company coaches or the Guangzhou Metro, to work. Those respondents who sought jobs through social networks tended to take slower transport modes, e.g., public buses or walking, to work. This latter group of respondents tended to take short-distance trips (Table 5.1). The social exclusion process involved the interaction between social networks and individual actions. The interaction produced new structural properties that included the unskilled and young respondents using their social networks to find jobs near their homes to cut transport costs. The respondents who were skilled and had high educational attainment tended to seek high-paying jobs in other districts in Guangzhou, and they drove or took the Guangzhou Metro to work. The results confirm the interaction between the social networks, educational attainment and job search actions of low-income local *hukou* respondents.

5.12 Summary of the Survey Results in Xiguan

This study investigated the social exclusion process in the historic inner city of Xiguan. The focus was on the interactions between structural factors and the individual actions of local *hukou* residents. The housing and economic reforms interacted with local residents' actions to produce new structural properties, which included the local residents earning low income and subdividing their houses into small rooms to accommodate more people as well as redundant workers trying to find jobs in the informal retail markets that were created by the revitalisation policies. The municipal government tried to redevelop the old houses that were deprived of kitchens, flush toilets and privacy.

Many of the local residents who were young or female tended to rent their flats and seek jobs in the retail markets in Xiguan. They showed a higher rate of residential mobility than the residents who owned their flats. Most of these renters had moved from other districts in Guangzhou to Xiguan to seek jobs through their social networks. Low-income women tended to find part-time jobs so that they could resolve the time-conflict commuting problems between their jobs and household responsibilities. These residents took short-distance trips and used slow transport modes. The residents earned a low income, particularly the women. Hence, they experienced social exclusion in terms of participating in distant and high-paying job opportunities. These residents also did not have sufficient time intervals to participate in recreational and social activities (Fig. 5.16).

The middle-aged local residents with higher educational attainment strong social networks tended to find higher-paying jobs in other districts and adopt faster transport modes to reach their workplaces. These groups of residents tended to own their flats and thus exhibited a low rate of residential mobility. They depended on social networks to find jobs in other districts. These groups also used faster transport modes and spent longer time travelling to work than young working residents did. In the face of travel congestion, outdated road systems and the inadequate provision of public bus services, the middle-aged residents had longer average travel times than the middle-income workers in Guangzhou. They were deprived of reasonable mobility to access employment (Fig. 5.16).

The last group of residents in Xiguan were the migrant workers, who were restricted by the *hukou* system from participating in the formal labour markets in the urban areas of Guangzhou. To survive in urban areas, they sought informal jobs, mostly through their social networks. In addition to restrictions from the *hukou* system itself, the rural migrants also faced stigmas due to their rural *hukou* status. Hence, these migrants had to develop an enclave on the eastern side of the neighbourhood to help one another find jobs and affordable housing. The population characteristics of migrants in Xiguan are slightly different from those of migrants in urban villages in Guangzhou.

A large share of the migrants came from other cities in Guangdong Province, of which Guangzhou is the capital city. These migrants could speak local languages and communicate with local residents. Thus, approximately 42% of them owned their apartments, and their average age was younger than that of the local residents. Most of them were young workers. Of all of the migrants, approximately 54% were single, and their average travel time was 24.9 min, while approximately 46% of them were married, and their average travel time was 26.6 min. Their intended goal was to find jobs in the informal markets in Xiguan.

Most of them sought jobs through their social networks. The migrant women faced time conflicts and commuting problems, and they thus sought retail jobs near their homes. Most of them adopted slow transport modes to reach work, such as walking, riding bicycles and taking public buses. They were excluded from participating in the mainstream activities of society, such as having social contact with the locals and reaching recreational and educational facilities (Figs. 5.16 and 5.17).

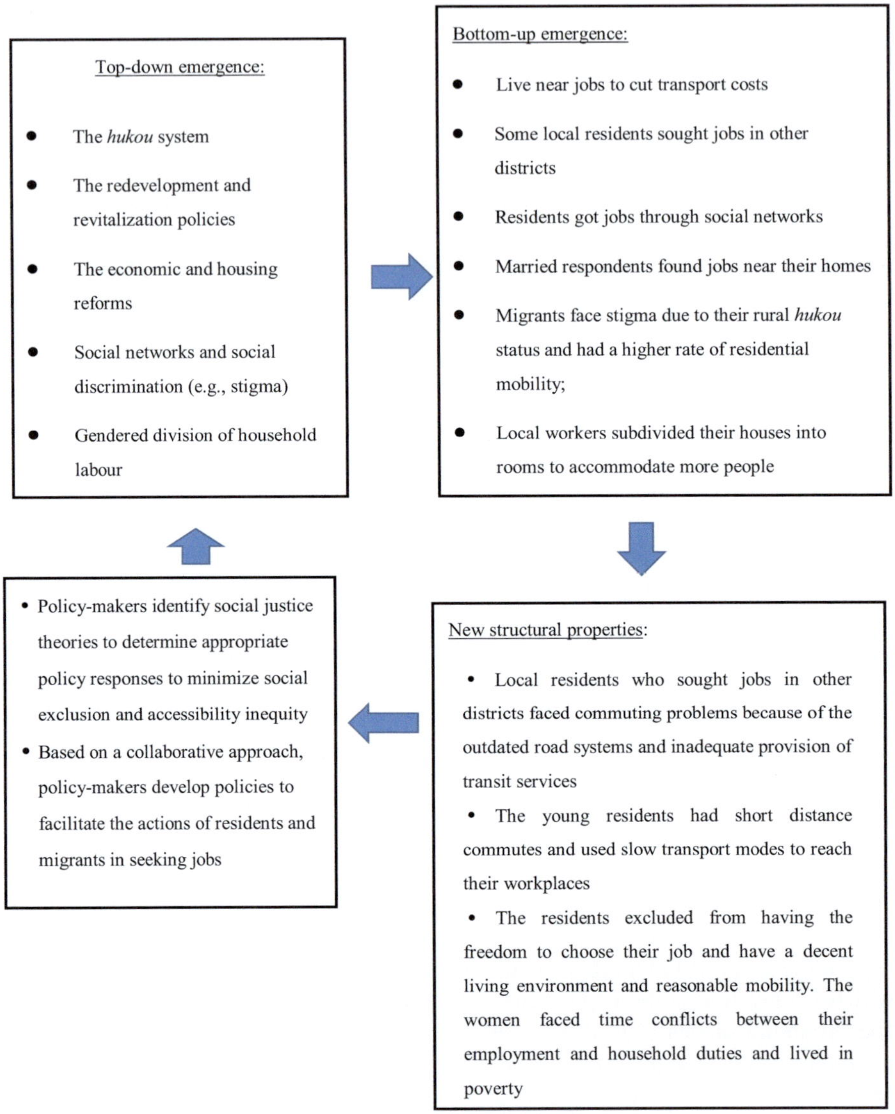

Fig. 5.17 Survey results: Social exclusion processes in Xiguan

5.13 Conclusion

The results of Chap. 5 confirm the second hypothesis of this book that the social exclusion process, which is taking place in the historic inner city of Xiguan in Guangzhou, produces new behaviours, poverty, accessibility inequity and capability deprivation. Social exclusion is a multi-dimensional phenomenon that shaped the investigation in Chap. 5 of low-income residents in Xiguan. One of the dimensions that causes the social exclusion of the low-income residents in Xiguan is poverty; the residents experience poverty because of the economic reforms, unemployment and their low educational attainment. The houses they live in are not only dilapidated but also deprived of flush toilets because of the housing reforms and the ineffective and inefficient urban redevelopment policies. Many low-income residents' work trips are constrained by outdated road systems, traffic congestion and the inadequate provision of transit services.

The second dimension that causes the social exclusion of the residents in Xiguan is the capability deprivation of the residents. The Guangzhou government does not provide the residents with the freedom, choice and opportunity to achieve well-being. The *hukou* system and economic reforms have deprived many of the residents of the capability to achieve well-being, resulting in their social exclusion. For example, the low-income residents are not only constrained by the *hukou* system and economic reforms, they also lack the freedom, choice and opportunity to exchange their labour power for adequate income and housing in Guangzhou. Many local residents do not have the freedom or opportunity to move out of the deprived inner city because they cannot afford to buy the commodity housing built by the private sector. Furthermore, the stigma and discrimination against rural migrant residents also causes the residents' capability deprivation. Rural migrant workers consistently face stigma from local residents in Xiguan. As a result, they experience discrimination and constraints in achieving well-being. The stigma against migrants excludes them from participating in certain types of retail jobs in the inner city and leads to them being socially excluded from mainstream society.

Finally, low-income residents face a dimension of accessibility inequity that causes exclusion. Influenced by unjust policies and social norms, many of the residents spatially restrict their jobs and social activities to the area of the deprived neighbourhood. The residents face accessibility problems in participating in social and recreational activities. Some local residents have to have longer commute times for their work trips than the middle-income workers in rich neighbourhoods. Some residents with low skills and low educational attainment have adopted slow transport modes to reach activities. They live and work in the deprived neighbourhood and are excluded from accessing distant job opportunities and social and recreational facilities, which are the mainstream activities of society.

Another case of accessibility inequity is that of working mothers in Xiguan. Because of the gendered division of household labour, many low-income working mothers have to take on the majority of the household responsibilities. As a result, the working mothers in Xiguan face time-conflict problems between their employment and household responsibilities. They do not have adequate time to travel long distances to seek jobs. Hence, they earn lower income and travel shorter distances on work trips than men. Some of the working women have to seek part-time jobs to resolve the jobs-housing time-conflict accessibility problems.

The mechanism of the social exclusion process is explained by structuration theory. Government decision-makers can minimise social exclusion by changing either the structure or agency. It is always easier to modify government policy to minimise social exclusion. The social exclusion process that is taking place in Xiguan indicates that several government policies, such as the economic and housing reforms and the *hukou* system, are incompatible with social justice. Rawls (1971) proposed the difference principle, which states that social and economic inequalities should be arranged so that they are to the greatest benefit of the least advantaged, such as the low-income local residents and migrants. The results of the investigations in this chapter indicate that the economic and housing reforms have excluded low-income workers from accessing decent homes, formal employment and good social relationships with migrants. Therefore, the economic and housing reforms are unjust policies and should be transformed to increase the social inclusion of low-income residents in Xiguan.

In addition, the capability approach suggested by Sen (1980, 1984, 1992) stresses that all people are entitled to have the freedom to achieve well-being and that the freedom to achieve well-being is to be understood in terms of people's capabilities, that is, people should have the freedom, choice and opportunity to obtain employment and housing and form social networks. Based on the capability approach, the municipal government should consider improving the capability of low-income local residents (Fig. 5.16). First, the municipal government should adjust its housing policy by providing more affordable housing for local residents in Xiguan so that they have the freedom and choice to select affordable housing and can live in decent homes that are in proximity to formal job opportunities. Second, the redevelopment of the old houses should be the responsibility of the municipal government, which should directly provide financial resources to redevelop all of the old houses in Xiguan, preserve the historic old buildings in accordance with the Chinese government's conservation policy, and compensate and resettle the deprived residents. Third, the narrow streets should be widened and the development of underground shopping streets should be considered to avoid conflicts between traffic and pedestrians. The concept of underground shopping streets has been widely adopted in cities in Japan, Australia and Singapore; the municipal government should build underground streets in Xiguan to avoid conflict between traffic and pedestrians.

The results of the surveys in this chapter indicate that migrant workers in Xiguan are prevented from accessing decent homes, formal employment and social welfare. Therefore, the *hukou* system should be transformed by the Chinese government to improve the accessibility and social inclusion of rural migrant workers. The migrant workers should be allowed to participate in formal labour markets, including government jobs, and to partake of the social welfare services provided by the municipal government.

In addition, government planning solutions need to be supplemented by bottom-up community participation. Decision-makers need to understand the attitudes and actions of the local residents and migrant workers in Xiguan to formulate appropriate policy responses to provide more freedom, choice and opportunity for low-income workers to achieve well-being (Fig. 5.16).

The social structure is capable of both constraining and enabling action (Giddens 1984). Healey (1997, 1998) proposes a collaborative approach for public participation in order to understand the individual actions and attitudes of low-income workers, which will help policy-makers modify unjust policies, enable action, and minimise social exclusion. The Xiguan District government should consult local residents about the redevelopment of old houses and improve the living environment of the residents. In addition, the local government could conduct interviews with rural migrants and local residents about their difficulties in finding jobs and housing in the inner city as well as about their actions to cope with over-crowded living environments and the *hukou* system. Chinese policy-makers should tolerate diversity and difference, the heterogeneity of values, and lifestyle oppositions in order to understand the residents' difficulties, skills and capabilities. When policy-makers understand the actions and attitudes of affected low-income workers, they can deliver appropriately just policies to tackle social exclusion in Xiguan.

References

Chen X, Stanton B, Kaljee LM, Fang X, Xiong Q, Lin D, Zhang L, Li X (2011) Social stigma, social capital reconstruction and rural migrants in urban China: a population health perspective. Hum Organ 70(1):22–32

Church A, Frost M, Sullivan K (2000) Transport and social exclusion in London. Transp Policy 7 (3):195–205

Dai D, Zhou C, Ye C (2016) Spatial-temporal characteristics and factors influencing commuting activities of middle-class residents in Guangzhou City, China. Chin Geogr Sci 26(3):410–428

Dokmeci V, Altunbas U, Yazgi B (2007) Revitalisation of the main street of a distinguished old neighbourhood in Istanbul. Eur Plan Stud 15(1):153–166

Fuchs C (2003) Structuration theory and self-organization. Syst Pract Action Res 16(2):133–167

Giddens A (1981) A contemporary critique of historical materialism. Vol. 1. Power, property and the state. Macmillan, London

Giddens A (1984) The constitution of society. Polity Press, Cambridge

Google Map (2018) Google map: Guangzhou 2018. http://maps.google.com/maps. Accessed 2 May 2018

Guangzhou Statistics Bureau (2017) Guangzhou statistical yearbook 2017, China Statistics Press, Guangzhou (in Chinese). http://210.72.4.52/gzStat1/chaxun/ndsj.jsp. Accessed 9 Sept 2018

Guangzhou Statistics Bureau (2018) The 6th Guangzhou population census in 2010, Guangzhou Municipal Government, Guangzhou (in Chinese). http://www.gzstats.gov.cn/pchb/dlcrkpc/. Accessed 1 Sept 2018

Healey P (1997) Collaborative planning: shaping places in fragmented societies. UBC Press, Vancouver

Healey P (1998) Collaborative planning in a stakeholder society. Town Plan Rev 69(1):1–21

Houston D (2005) Employability, skills mismatch and spatial mismatch in metropolitan labour markets. Urban Stud 42(2):221–243

Ihlanfeldt KR, Sjoquist DL (1998) The spatial mismatch hypothesis: a review of recent studies and their implications for welfare reform. Hous Policy Debate 9(4):849–892

Kain JF (1968) Housing segregation, negro employment, and metropolitan decentralization. Q J Econ 82(2):175–197

Li X, Stanton B, Fang X, Lin D (2006) Social stigma and mental health among rural-to-urban migrants in China: A conceptual framework and future research needs. World Health Popul 8 (3):14–31

Lin GCS (2015) The redevelopment of China's construction land: practising land property rights in cities through renewals. China Q 224:865–887

Litman T (2017) Evaluating accessibility for transport planning, Victoria Transport Policy Institute. http://vtpi.org/access.pdf. Accessed 14 June 2018

Liu YT, Wu FL, He SJ (2008) The making of the new urban poor in transitional China: market versus institutionally based exclusion. Urban Geogr 29(8):811–834

May DM (2004) The interplay of three established–outsider figurations in a deprived inner-city neighbourhood. Urban Stud 41(11):2159–2179

McCabe BJ, Ellen IG (2016) Does preservation accelerate neighborhood change? J Am Plan Assoc 82(2):134–146

Parks V (2004) Access to work: the effects of spatial and social accessibility on unemployment for native-born black and immigrant women in Los Angeles. Econ Geogr 80(2):119–220

Porter ME (2016) Inner-city economic development: learnings from 20 years of research and practice. Econ Dev Q 30(2):105–116

Rauws W, De Roo G, Zhang S (2016) Self-organisation and spatial planning: an editorial introduction. Town Plan Rev 87(3):241–251

Rawls J (1971) A theory of justice. Oxford University Press, Oxford

Sen A (1980) Equality of what? Choice, welfare and measurement (Cambridge, MA: Harvard University Press), 1997 [Originally published in The Tanner Lectures on Human Values, Vol. I, by the University of Utah Press and Cambridge University Press, 1980]

Sen A (1984) Resources, values and development. Oxford, Basil Blackwell

Sen A (1992) Inequality reexamined. Oxford University Press, Oxford

Shen J (2017) Stuck in the suburbs? Socio-spatial exclusion of migrants in Shanghai. Cities 60:428–435

Tan XH, Altrock U (2016) Struggling for an adaptive strategy? Discourse analysis of urban regeneration processes – a case study of Enning road in Guangzhou City. Habitat Int 56:245–257

Wang YP (2000) Housing reform and its impacts on the urban poor in China. Hous Stud 15 (6):845–864

Wang Z, Zhang FZ, Wu FL (2017) Social trust between rural migrants and urban locals in China – exploring the effects of residential diversity and neighbourhood deprivation. Popul Space Place 23:1–15

Wu N, Zhao SC, Zhang Q (2016) A study on the determinants of private car ownership in China: findings from the panel data. Transp Res A 85:186–195

Yuan Y, Wu FL (2014) The development of the index of multiple deprivations from small-area population census in the city of Guangzhou. PRC Habitat Int 41:142–149

Yuan Y, Wu FL, Xu XQ (2011) Multiple deprivations in transitional Chinese cities: a case study of Guangzhou. Urban Aff Rev 47(5):721–736

Zacharias J (2012) Resisting motorization in Guangzhou. Habitat Int 36:93–100

Zhang C, Li X (2016) Urban redevelopment as multi-scalar planning and contestation: the case of Enning road project in Guangzhou, China. Habitat Int 56:157–165

Zhou SH, Deng LF, Kwan MP, Yan RG (2015) Social and spatial differentiation of high and low income groups' out-of-home activities in Guangzhou China. Cities 45:81–90

Chapter 6
Suburbanisation of Poverty in Shanghai and Declining Ridership of Rail Services

6.1 Introduction

Shanghai is the largest city in China in terms of population and is an important economic, financial, trade and shipping centre in China. In 2016, the total population size of Shanghai reached 24.2 million, of which 9.72 million were low-income rural migrant workers (Shanghai Statistics Bureau 2017b). Shanghai's annual GDP growth rate was initially 7.8 in 2013, reaching 6.9% in 2017 (Shanghai Statistics Bureau 2017b). However, following its remarkable economic growth, Shanghai has experienced rapid economic polarisation and the suburbanisation of poverty, which have resulted in the exclusion of low-income workers from participating in mainstream activities of society, such as formal employment, decent housing, adequate income and social activities.

Sassen (2001) proposes that global cities such as Shanghai have emerging trends of economic polarisation, where dual markets exist for both professionals and low-income workers in various industry sectors at the same time. Polarisation describes the degree to which a society is segregated into groups. In this chapter, the focus is on the spatial segregation of low-income workers and professionals during the process of suburbanisation.

6.1.1 Suburbanisation of Poverty in the Cities of Western Countries

The suburbanisation of poverty is a kind of social exclusion process or self-organisation process. Under globalisation, the fundamental shifts in urban economies, land–use policies and labour markets have caused a number of world cities to gradually shift away from the pattern of suburbanisation of the large cities in the

J. C.-y. Lau, *Self-Organisation Shapes Travel Behaviours and Social Exclusion in Deprived Urban Neighbourhoods of China*,
https://doi.org/10.1007/978-981-15-2252-9_6

U. S., which was investigated in Chap. 5. The suburbanisation of poverty emerged in large cities in Western countries, and the process displaces low-income workers from urban areas to suburbs because of rapid social and economic development (Bailey and Minton 2018). In London and other cities in Western countries, the wealthy living have recently begun living in urban areas and the poor have begun living in suburbs. In addition, some studies in Australia have found that there has been a trend towards the marked suburbanisation of low-income households and homelessness in Australian cities (O'Donnel 2018; Randolph and Tice 2017). In Holland, the outward expansion of gentrification away from the inner city into other neighbourhoods has led to outward shifts of poverty into the suburbs in the urban periphery, particularly in Rotterdam and Amsterdam. The high land prices in the urban areas price out low- and middle-income households from the owner-occupied sector. Many low-to-middle income households are increasingly moving to urban regions surrounding cities, particularly to higher density satellite towns (Hochstenbach and Musterd 2018).

The suburbanisation of poverty in large cities involves a cycle of structure and agency (Giddens 1984). The structural factors that further the suburbanisation of poverty in European cities include urban redevelopment, economic growth, the development of fast public transit systems, high land prices in city centres and the structural change of the economy, which has led higher income groups to move into central urban areas (Bailey and Minton 2018). Regarding individual decisions that add to the suburbanisation of cities, many low-income people exercise choice and relocate to suburban areas because of the relatively low land price and house rents there, as well as because of work-place proximity; choice and proximity have had the most significant influence on individual decisions to move to the urban periphery.

Many relocated workers can find cheaper private rents or welfare housing in suburban areas. Some want to live near the job opportunities in the new towns. Some displaced workers decide based on the trade-off between a large living environment and access to employment. They choose to live in a large residence in the suburbs and travel long distances from their homes to employment opportunities in urban centres. Thus, the suburbanisation of poverty in European cities is a social exclusion process that excludes low-income workers from accessing major public services, social networks and employment opportunities. Few studies have investigated this issue in developing cities; this chapter attempts to fill this research gap by examining the issue of the suburbanisation of poverty in Shanghai by means of the social exclusion process and self-organisation approach.

6.1.2 The Suburbanisation of Poverty in Shanghai

Shanghai has also experienced an overarching trend of the suburbanisation of poverty in the inner suburbs and surrounding regions. However, the suburbanisation patterns in Shanghai are significantly different from the patterns of suburbanisation in Western cities. The Shanghai municipal government has carried out some suburbanisation policies that do not appear in Western countries. For example, in order to produce cheap and competitive manufacturing products to compete in global markets, the Shanghai municipal government follows the dual labour market policy. Based on the *hukou* system, formal jobs in Shanghai are nearly exclusively reserved for residents with a Shanghai *hukou,* while most rural *hukou* migrants are channelled into informal labour markets (Feng and Chen 2018; Wan and Wang 2015). To find informal jobs and survive in urban areas, rural migrants choose to gather in the inner suburbs to find informal manufacturing jobs or to borrow start-up cash for developing small retail businesses. They live and work near industrial parks and migrant enclaves. This is a phenomenon of social exclusion. The rural migrant workers live in poverty because of their low income; they also experience a lack of freedom and choice in exchanging their labour power for food and housing.

Furthermore, the Shanghai municipal government finances its infrastructure development and economic growth using land sales. This land finance policy contributes to increasing real estate prices, resulting in housing in the downtown becoming unaffordable for low-income workers (Wu et al. 2015; Peterson 2006). Many low-income workers are unwilling to relocate to the suburbs. However, the drastic policies of the municipal government and privately funded redevelopment policies force low-income workers to move to the suburbs. Approximately 22.2% of the local *hukou* households relocated from the city centre to suburban areas between 1995 and 2011 during this social exclusion process (Chang and Chen 2013).

To obtain adequate land to build industrial parks and to build public housing for displaced workers, the Shanghai government acquires farmland that is collectively owned by villagers in suburban areas. The farmland that is acquired is always located in the areas are far from the main public transit networks. As a result, many industrial parks and resettlement housing are dispersedly built in suburban areas. Many migrant workers and displaced workers find it difficult to take public transit services to their jobs as a result of the urban sprawl.

To provide a good connection between the suburban area and the downtown, the Shanghai government built the Shanghai Metro, an excellent rail transit system. As of 2016, the metro system included 17 lines and 395 stations, and the total length of the rail transit network added up to 673 km (Pan et al. 2017). However, the actual travel behaviours of the residents in the suburbs differs slightly from that which was forecast using the conventional urban transport planning model and the design guidelines of transit-oriented development (TOD) and which was used to plan the current rail system (Wang et al. 2012; Cervero and Day 2008). The results of one study found that the Shanghai Metro fails to attract adequate travellers from new towns to take long-distance trips. Consequently, the Shanghai Metro faces insufficient rail patronage that negatively influences the cost-effectiveness of the rail transit services (Guan and Xu 2018; Pan et al. 2011).

This book argues that the process of suburbanisation produced new travel behaviours of low-income suburban workers. The new travel behaviours include taking short-distance trips and adopting slow transport modes to reach activities. Hence, these workers do not need to use rail transit services for their daily trips. For example, in 2016, migrants accounted for approximately 40.5% of the total population in Shanghai, and most of them lived in the inner suburbs (Shanghai Statistics Bureau 2017a). Because of the expropriation of collectively owned land and the sprawling of industrial zones in the suburbs, most of the migrants, who live near industrial job opportunities, live far away from the major rail lines in the suburbs. The factors of dispersed jobs and housing exclude migrants from using rail services for long-distance trips (Guan and Xu 2018; Pan et al. 2011).

6.1.3 Self-Organisation Approach and Social Exclusion

The suburbanisation of poverty in Shanghai can be investigated in terms of the social exclusion process and self-organisation approach, in which government policies and social norms influence the suburbanisation of people and industries, but at the same time, the policies are reproduced and transformed by relocated local residents and migrants in their routine practices. The process produces new structural properties, which cause an overarching trend of the suburbanisation of poverty in the urban peripheries. The process excludes displaced workers from accessing employment, housing, and recreational and social activities (Fuchs 2003; Giddens 1981, 1984). For example, the land finance and redevelopment policies interact with the actions of the low-income residents to produce new structural properties, which include the relocation of low-income local residents to suburban areas and their exclusion from accessing employment and non-work activities (Tian et al. 2017a, b, c). Another moving group is rural migrant workers. The interactions between the *hukou* system and the actions of migrants have produced new structural properties, which have led to the exclusion of migrants from accessing formal jobs and welfare benefits. They choose to move the inner suburbs to find jobs and make a living.

Furthermore, the self-organisation process can be used as a tool to investigate the multi-dimensional phenomenon of social exclusion, which is also a kind of self-organisation process. Social exclusion is defined as low-income workers' inability to achieve well-being due to resource deprivation and a lack of freedom, choice and capability to participate in the mainstream activities of society. This chapter argues that suburbanisation policies cause relocated low-income residents and migrants to face social exclusion. Suburbanisation policies cause social inequity. Low-income rural migrants experience accessibility problems in meeting basic needs because their workplaces and homes are far away from the main transit networks due to the land acquisition policy. The displaced local workers also face accessibility problems because the relocation public housing estates are dispersed within the inner suburbs due to urban sprawl and the land finance and urban redevelopment policies.

Low-income workers used their individual decisions and actions to transform the policies through their routine practices. For example, many rural migrants develop enclaves in the new towns that have been transformed into business and retail centres, such as the Zuibaichi Park neighbourhood in the Songjiang District. The hypothesis of Chap. 6 is that the process of the suburbanisation of poverty in

Shanghai has produced new travel behaviours, social exclusion, accessibility problems and the declining ridership of rail services.

In this chapter, the author poses the following three research questions: First, how does the process of suburbanisation in Shanghai produce social exclusion and commuting problems? Second, could the built environment and transport system integration alone produce sufficient ridership to support cost-effective rail transit services in the suburbs of Shanghai? Third, which policies should the government implement to minimise social exclusion and increase rail transit ridership? Furthermore, the author expects that the investigation of the above three research questions will shed light on the research related to urban transport planning, social exclusion and self-organisation.

The main body of this chapter is organised into five sections. The first introduces the Master Plan (1999–2020) that guided the suburbanisation of the population and industries in Shanghai. The next section investigates the cycle of structures (e.g., urban redevelopment, housing and land finance policies) and agencies that have produced the suburbanisation of poverty, new travel patterns and social exclusion. The third section investigates the cycle of structures (e.g., the *hukou* system, social networks, industrial job suburbanisation, and the demolition of urban villages) and rural migrants' individual actions that produce migrant enclaves, social exclusion, new job opportunities and new travel behaviours that result in the declining ridership of rail services, which fails to support cost-effective rail transit services. The fourth section presents a route test on the performance of the Shanghai Metro. In addition, an on-site survey was conducted to evaluate whether the built environment around the three chosen rail stations is developed on the same level as the TOD design guidelines. The test and survey results confirm that the suburbanisation process of poverty is the root cause of the declining ridership of rail services. The final section presents the conclusion of this chapter.

6.2 Shanghai Master Plan (1999–2020) and Social Exclusion Process

Two decades ago, the decision-makers of the Shanghai municipal government developed a suburbanisation plan when they observed the population increase and the rapid economic growth of the city. The policymakers introduced the Shanghai Master Plan (1999–2020), which proposed the '1-9-6-6 Model'. The plan intended to suburbanise a large proportion of the population, moving them from the city centre to the suburbs in order to shape a multi-centred structure. The model was represented by 1 central city that included all urban districts, 9 satellite cities (the Baoshan, Jiading, Qingpu, Songjiang, Minhang, Fengxian, Nanqiao, Jinshan, Liangang districts), approximately 60 new independent towns around the 9 districts, and 600 villages in the remote suburbs (Yang et al. 2015; Shen 2011; Pan 2006). The satellite cities developed major industries, commodity housing and key urban

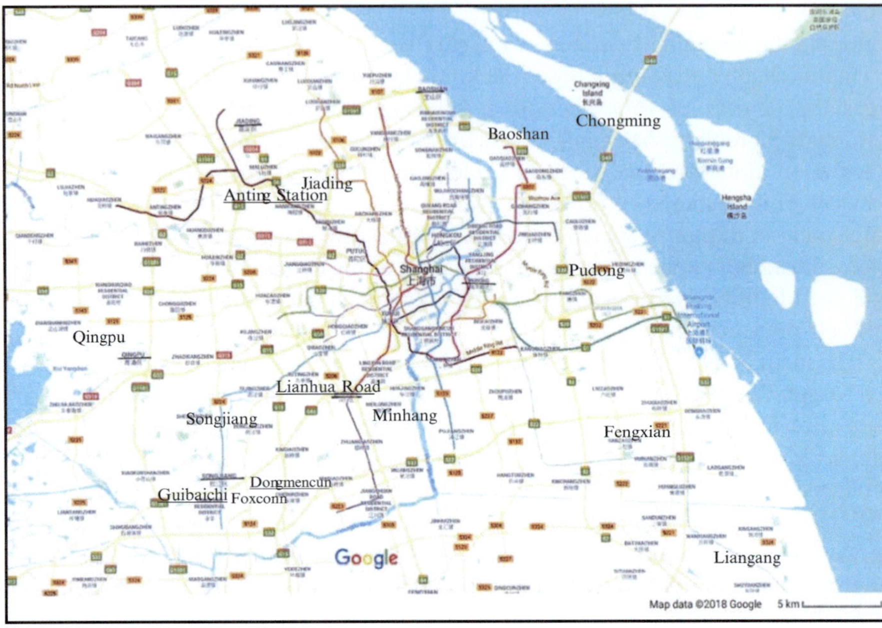

Fig. 6.1 The new towns and Shanghai rail transit system. Source: Modified from Google Maps (2019)

infrastructure projects (Fig. 6.1). Nine Western-style towns were built in these satellite cities to attract high-income residents and expats. Each town had its own architecture and planning theme that followed Western traditions, such as a British town in the Songjiang District and a town with German-style architecture in the Jiading District. The low-income residents were resettled into affordable housing. In addition, manufacturing industries were suburbanised to industrial parks in the new towns. Rail transit systems were built to provide a good connection between the new towns and the city centre (Fig. 6.1).

The social exclusion process, in which the individual actions of relocated local residents and rural migrants interact with suburbanisation policies, produces new structural properties, including the concentration of a low-income population in the suburbs. The Shanghai Master Plan (1999–2020) ignores the decisions and actions of relocated low-income residents and rural migrants. As a result, the plan had led to the discrepancies between the predicted land-use plan and the actual spatial distribution of the low-income workers; for example, a large proportion of the migrant population is concentrated in the inner suburbs, and they contribute to the suburbanisation of poverty.

In 2016, the total permanent population in Shanghai was 24.19 million, of which approximately 9.7 million were rural migrants. Among all the rural migrants, approximately 68.6% lived in the inner suburbs, which is the band between 10 and 20 km from the city centre. The inner suburbs include the Pudong, Baoshan,

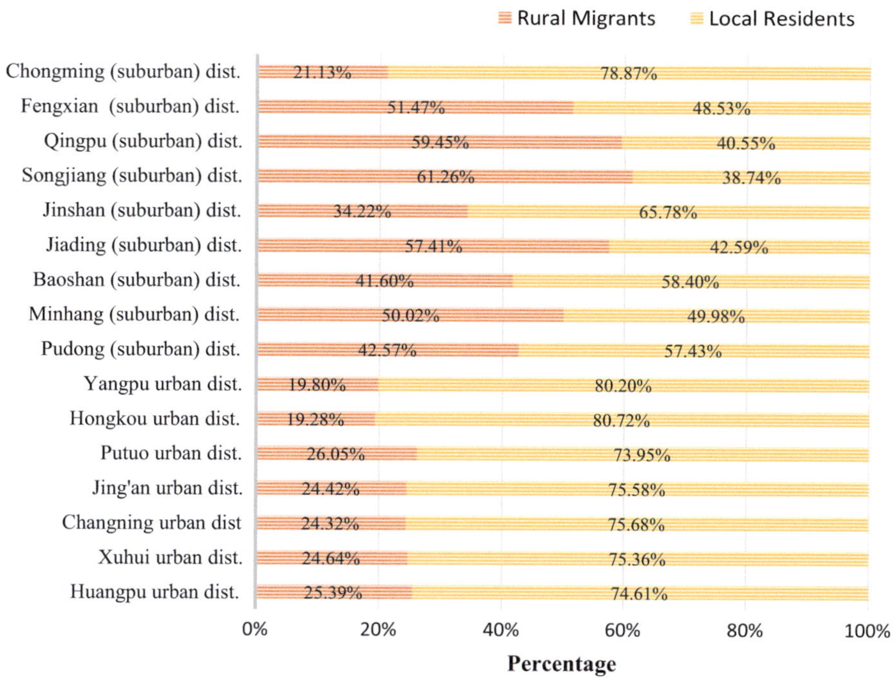

Fig. 6.2 Population distribution between migrants and local residents by district in Shanghai. Source: Shanghai Statistics Bureau (2017a)

Jiading, Songjiang and Minhang districts (Figs. 6.1 and 6.2). In some of the new towns and districts, such as Minhang, Jiading, and Songjiang, the number of migrant workers has overtaken the number of local *hukou* residents (Fig. 6.2) (Shanghai Statistics Bureau 2017a).

The Master Plan (1999–2020) estimated the steady growth of the population. The local residents of Shanghai were predicted to increase from 14.57 million in 1998 to 15 million in 2010 and to reach 16 million in 2020. The local *hukou* population in Shanghai remained stable, which was in line with the estimations of the Master Plan (He 2013). However, the actual population reached 24.2 million in 2018 (Shanghai Statistics Bureau 2017a). Clearly, the Master Plan failed to predict the actions rural migrants would take to find jobs and affordable housing in the inner suburbs, and their actions produced the phenomenon of the suburbanisation of poverty (Fig. 6.2).

6.3 Social Exclusion Process Shapes the Suburbanisation in Shanghai

The self-organisation process can be used as a tool to investigate social exclusion. Social exclusion is a process in which unjust policies prevent low-income workers from having the freedom, choice and opportunity to achieve well-being. The workers have been excluded from accessing mainstream activities during the process of suburbanisation in Shanghai (Stanely et al. 2019). In other words, the policies interact with the actions of low-income local residents and migrant workers to produce new structural properties that lead to the exclusion of low-income workers from accessing employment and social and recreational activities. The structural factors that influence the social exclusion of low-income workers include the following unjust government policies (Tian et al. 2017a, b, c; Yang et al. 2015; He and Wu 2007) (Fig. 6.3):

1. The government adopted land financing and privately funded property-led redevelopment policies, which make the land prices in the city centre unaffordable for the affected low-income residents, resulting in the suburbanisation of poverty (Shen 2011).
2. The municipal government demolished urban villages and TVEs, leading to the sprawling of migrant enclaves in the inner suburbs;
3. The municipal government acquired collectively owned farmland for building industrial zones and affordable housing, which has led to urban sprawl;
4. The *hukou* system restricts migrants (who represent approximately 40.5% of the total population) from finding formal jobs and housing in the city centre; and
5. Public transit services fail to cover all housing and employment locations in suburbs because of urban sprawl.

The above variables interact with the actions of the relocated local residents and migrant workers to form a multi-dimensional phenomenon of social exclusion. In addition, the social exclusion process can be investigated in the form of a two-stage interaction, top-down emergence and bottom-up emergence (Fuchs 2003). Top-down emergence refers to structures (e.g., rules and resources) that constrain and enable practices, such as the *hukou* system. Bottom-up emergence is the agency of low-income workers. For example, a large share of rural migrants, who are excluded by the *hukou* system from accessing formal jobs, move from the city centre to the inner suburbs and find informal industrial and retail jobs. They live close to job opportunities and develop migrant enclaves in the inner suburbs. In addition, a high number of relocated local residents involuntarily move to relocation housing in suburban areas as a result of the property-led urban redevelopment and land finance policies (Guan and Xu 2018; Yang et al. 2015; He and Wu 2007).

The relocated local workers try to find new employment near their homes, but there are not adequate job opportunities for them because of the problems of urban sprawl. The younger relocated local workers do not want to find new jobs; thus, they have to travel long distances to reach their workplaces in the city centre. Many

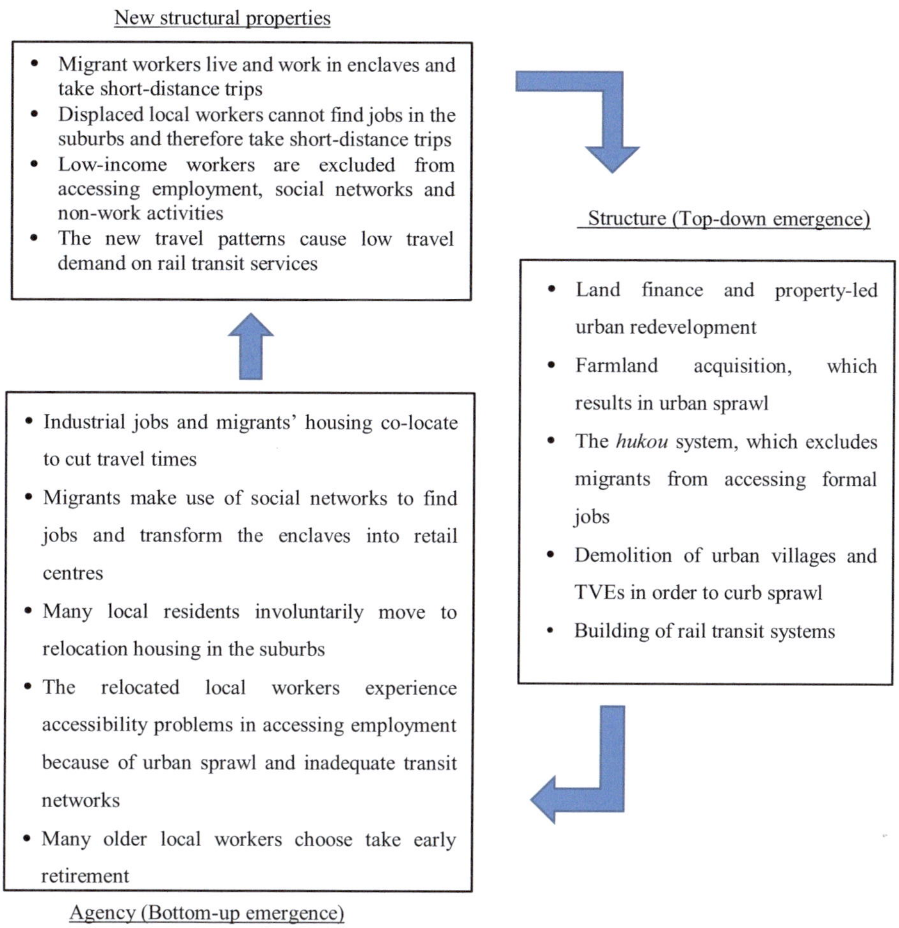

Fig. 6.3 Social exclusion process shapes the suburbanisation in Shanghai

relocated low-income and middle-aged workers choose to take early retirement and apply for social welfare. The interaction between policies and agencies produces new structural properties. The properties include the concentration of poverty in the suburban areas and a large share of migrants and relocated local residents taking short-distance trips and using slow transport modes to reach activities. Most of the relocated low-income local workers experience accessibility problems in accessing employment, social networks and recreational activities. In addition, the new travel behaviours of the relocated local residents and migrants has led to the declining ridership of rail transit, which is insufficient to support cost-effective rail transit services (Fig. 6.3).

6.4 Land Finance Policy Makes Land Prices in the City Centre Unaffordable for the Affected Local Residents

Land finance policy interacts with low-income local workers and produces new structural properties, which include the land prices in the city centre becoming unaffordable for many low-income residents who then choose to relocate from dilapidated urban neighbourhoods to inaccessible suburban areas. The self-organisation process began during the housing reforms of the 1990s. During that time, housing gradually changed from being welfare provided by *danweis* to being a commodity housing market. This move was associated with the development of a land leasing system and real estate businesses. In 1990, the State Council confirmed that land-use rights could be conveyed from local governments to private property developers, and this approval marked the establishment of an urban land leasing system. As a result, land sales became an important way for the Shanghai municipal government to finance its investments in infrastructure. The improved transport infrastructure helped improve the accessibility of the city centre and raised land values to attract more investments. This type of fiscal revenue strategy was referred to as land finance (Wu et al. 2015). In addition, Shanghai senior government officials' performance is evaluated through their promotion of economic growth. They usually use land sales to finance city development and economic growth (Wu et al. 2015). In 2014, the revenue for land sales accounted for approximately 27% of the total revenue in Shanghai (OECD 2017).

Tian et al. (2017a, b, c) find that the average price of commodity houses has the strongest relationship with the expansion of urban settlement, followed by urban population size, the income of urban residents, investment in real estate industry, and GDP of the tertiary industry. In addition, housing prices in Shanghai are severely out of proportion to the incomes of most people. In 2016, an average 80 m^2 apartment within Shanghai's Inner Ring Road was approximately RMB 6,113,400 (US $886,000), which was approximately 69 times the average annual income in the city (Shepard 2016). The land finance strategy has resulted in unaffordable urban housing prices for many low-income residents.

Alonso (1964) proposed a bid-rent function that shapes the trade-off between accessibility and land price. Urban land with higher accessibility can be sold at a higher land price than the land in the suburbs. For example, studies have found that a community's geographic proximity to a Shanghai Metro station tends to correlate positively with the average asking rent of apartments, indicating a significant rent premium for transit proximity (Li et al. 2019; Wang et al. 2016). As a result, housing prices and accessibility play important roles in the 'sorting' of Shanghai's socio-spatial structure and the suburbanisation of poverty. The increasing housing prices in Shanghai have priced out low-income workers from old urban neighbourhoods (Yang et al. 2015; He and Wu 2007). In 2008, a study that investigated relocated local residents in three suburban neighbourhoods found that nearly 42% of the respondents were involuntary movers. These respondents were mostly low-income workers, and they lived in resettlement housing that was not near main rail transit networks. They lived far away from their work, social and community connections, which were mostly located downtown. The inadequate public transit provision also hindered them in accessing local amenities such as shops, restaurants, schools and healthcare facilities (Day 2013). The self-organisation process produces new structural properties, such as the inaccessibility of employment and social networks. In addition, the suburbanisation process excludes low-income workers with Shanghai *hukou* from participating in the mainstream activities of society.

6.5 Redevelopment Policy Provides Inadequate Compensation for the Affected Local Residents Moving to New Homes in the City Centre

The interaction between the property-led redevelopment policy and the actions of the low-income workers with Shanghai *hukou* status has produced new structural properties, including the redevelopment policy that has provided inadequate compensation for the affected residents to buy new urban housing. The self-organisation process started in the 1990s when the municipal government implemented rapid and massive urban redevelopment. The government intended to redevelop the old downtown areas and demolish the dilapidated housing. For example, the government adopted the '365 scheme' (i.e., it redeveloped 365 ha of old and dilapidated housing). The policy of using the privately funded and property-led redevelopment added to the effects of urban development (Fig. 6.3). An example of the property-led redevelopment was the Taipingqiao Project. In 1996, the Hong Kong Shui On Group signed a cooperative agreement with the Luwan District government for the Taipingqiao Redevelopment Project. The affected residents were resettled and compensated, and the cleared land space was leased out. The Shui On Group invested 25 billion RMB (US$3.6 billion) and was responsible for demolishing the old buildings and replacing them with high-rise modern structures, such as the famous Xintiandi Shopping Centre (He and Wu 2005, 2007).

As part of the government's property-led redevelopment approach, the compensation that the affected residents received from the demolition of their old inner-city homes, was not enough to buy new apartments in the inner city areas, and the actions of the affected residents involved moving to the suburbs. Many of the relocated local residents failed to find new jobs in the suburban areas, and they chose to take early retirement and apply for social welfare (Guan and Xu 2018; Chang and Chen 2013). Thus, the property-led redevelopment interacted with the actions of the low-income workers and produced new structural properties, which led to the exclusion of the affected residents from accessing employment opportunities and social networks. In addition, the interests of the displaced poor renters in the older urban neighbourhoods were largely ignored, and many of the renters in the overcrowded dwellings were migrant workers. Between 1995 and 2011, approximately 1,159,899 households with Shanghai *hukou* were displaced from the downtown area to suburban areas; in other words, approximately 22.2% of local *hukou* households were relocated (Chang and Chen 2013). In 2011, approximately 30 large-scale affordable housing estates were developed for relocated local residents, such as Pujiang in the Minhang District and Sijing in the Songjiang District (Tongji University 2011). The interaction between the property-led redevelopment policy and the actions of low-income workers has been one of the main drivers of the exclusion of low-income workers from accessing the mainstream activities of society and of the concentration of poverty in suburban areas.

6.6 New Travel Behaviour of the Relocated Local Residents

The interactions between the suburbanisation policies and the actions of the relocated local *hukou* residents produce new travel behaviours. Many of the relocated local residents have moved to the relocation housing estates. Most of the residents have low skill and educational levels, and they tend to not have agricultural jobs (Zhou and Logan 2005). A large number of elderly people have been displaced from central Shanghai to the relocation housing estates. The displaced low-income residents in suburban areas are less likely to use their neighbours to access job information than the residents in the city centre. The neighbourhood effect contributes to the high unemployment rate of the displaced local residents.

Shen (2011) finds that since 2000, many people have left Shanghai and bought houses in the suburbs. In most cases, they have had a lump sum compensation payment of approximately two to three million RMB, while a flat in Shanghai cost only half that. Therefore, many families consider moving to the suburbs to be a good idea, particularly for retirees who do not have much to do in the central areas.

Many young people move to the suburbs because the housing in suburban areas is affordable (Shen 2011). Their jobs are located in the city centre, and some travel long distances from their homes to the city centre each day. Some of them make individual decisions and take individual actions to transform the relocation policies by leasing out their relocation housing to migrant workers. They rent or buy private flats in the city centre so that they can reduce the travel times of work trips.

Day and Cervero (2010) find that approximately 42% of relocated local residents have made involuntary moves to suburban areas because of government urban redevelopment policies. They experience increased travel times and decreased accessibility of regional jobs. These relocated local residents use their individual decisions and actions to interact with the suburbanisation policies and produce new structural properties, which include new travel behaviours. Yuan and Lin (2013) found that approximately 25% of the relocated commuters change jobs after such a so that their present workplaces are near their homes, and their average travel time is 35 min. They take public transit busses (approximately 35%) or ride e-bikes (approximately 65%) to work. A new travel behaviour of another group is that approximately 75% of the relocated local residents have failed to find new jobs in the suburbs. Some choose to take early retirement, and some choose to travel long distances from their homes in the suburbs to their workplaces in the city centre; their average travel time is 62.2 min. The rail transit networks are located far away from the relocation housing estates, and inadequate feeding bus services contribute to the long travel times of the workers.

Guan and Xu (2018) find that even though some relocation housing estates are located near rail lines, the relocated local residents experience accessibility problems. They choose among four neighbourhoods in the suburbs. The first is the Gucun neighbourhood in the Baoshan District, which is located near line 7 of the Shanghai Metro. The second is the Jiangqiao neighbourhood in the Jiading District, which is located near line 13 of the Shanghai Metro. The third is the Sijing neighbourhood in the Songjiang District, which is located near line 9. The last is the Pujiang neighbourhood in the Minhang District, which is located near line 8. In the survey, approximately 4800 relocated local residents provided valid results.

Many of the relocated workers felt that the transit service networks could not cover all of the housing estates in the new towns; thus, they were unable to take public transit to the jobs in the city centre. In other words, they experienced problems with unmet travel needs. In the face of jobs-housing spatial mismatch commuting problems, a high number of local *hukou* relocatees chose to take early retirement and apply for social welfare. For example, approximately 65% of the relocatees did not work, and 61% of them had an education level of junior secondary or less. Over 46% of them earned less than RMB 3000 per month in 2012. Concerning the transport mode choices for reaching work, approximately 39% of the relocated local residents took the bus, 26% took rail transit, 14% rode e-bikes, 5% walked, and only 5% drove to work. The younger relocatees tended to find jobs in the city centre, and they commuted longer travel distances and paid higher fares.

Compared to the rural migrant workers in the suburbs, the local *hukou* relocatees were older and poorer and had a higher mode share of riding the bus and lower mode share of driving cars. For example, approximately 66% of the local hukou relocatees were in the 50–70-year-old age group, and approximately 82% of them had Shanghai *hukou*. In addition, many relocatees experienced accessibility problems in reaching non-work activities, such as medical services and recreational facilities (Li et al. 2019; Guan and Xu 2018). Therefore, research evidence indicates that the self-organisation process, or the interaction between urban redevelopment policies and the actions of relocated local workers, has produced new structural properties, which include the relocated local workers generating short-distance trips on average and the exclusion of these workers from accessing the social networks and employment that are located in the city centre. The new travel behaviours of the relocated workers contribute to the low level of rail transit ridership that is insufficient to support cost-effective rail transit services.

6.7 Acquisition of Collectively Owned Farmland Produces Urban Sprawl

In Shanghai, urban expansion has taken place most rapidly in the peri-urban areas, and land use in these areas has experienced significant changes and restructuring. From 1990 to 2009, the area of non-agricultural land in the peri-urban areas of Shanghai increased 79,119 ha, which is even larger than the city centre at 66,000 ha. In terms of the urban expansion, industrial land was predominant in 1997, and the proportion of industrial land within all non-agricultural land in the peri-urban areas was 73.1%. However, the share of industrial land decreased to 55.0% in 2009 (Tian et al. 2017a, b, c). In the process of suburbanisation, the municipal government needed to acquire more and more collectively owned farmland in order to satisfy the need for the construction and development of economic and technological development zones (ETDZs).

There are two types of land that can be owned. The first is urban land in the city centre, which is owned by the municipal government. Urban land users are permitted to obtain land-use rights by bidding and through auctions and listings. The other type is collectively owned rural farmland in suburban areas. The development rights for rural land are strictly limited to farmers' houses, township and village enterprise (TVE) sites, and public facilities. The sale transfer or lease of collectively owned land for other non-agricultural uses is forbidden. However, the Shanghai municipal government has the power to acquire land from farmers and conduct land development and then to convey the land-use rights to private property developers.

Land price and government policy determine the location of the industrial zones in the suburbs (Zhang et al. 2018). First, planning policies boost enterprises located in areas planned for industrial land use. Second, the land price also determines the location of the industrial zone. In general, when a land resource becomes inadequate in the city centre, manufacturing enterprises will relocate to suburban areas with abundant land. That is, the market force discourages manufacturing enterprises from aggregating in the high-cost city centre and encourages them to relocate in the collectively owned farmland in the suburbs with lower land prices. Manufacturing enterprises prefer low land costs, and the land spaces that are converted from collectively owned farmland are located farther from roads and rail transit stations than the settlements that are developed on urban land (Zhang et al. 2018).

The results of one study found that industrial land sprawl occurs mainly in the form of leapfrogging. In other words, industrial land exhibits an inverted U-shaped trend of sprawling with increasing distance from the city centre. Most factories are concentrated in areas that are 25–55 km from the urban centre. In addition, ETDZs have mainly been developed in the Pudong, Baoshan, Jiading and Songjiang districts (You 2016). Since many rural migrants rent apartments and live near industrial job opportunities, the distribution of their homes follows the sprawling patterns of industrial locations in the inner suburbs.

In addition, urban sprawl features numerous TVEs that are scattered in the inner suburbs. In 2014, the area of TVEs accounted for 23.8% of all industrial land in

Shanghai, but its contribution to industrial output was less than 10%. Thus, the impact of TVEs on economic growth was much less significant than the state-led suburbanisation of industries (Tian et al. 2017a, b, c). The industrial zones in the suburban districts gradually merged with the scattered industrial spots, such as TVEs in urban villages (Fig. 6.3). In the process of land acquisition of urban villages and TVEs, the government compensated displaced villagers at a rate far below fair market value. The price difference between land-transfer fees and compensation turned into extra-budgetary revenue for the municipal government (Liu et al. 2018a, b; Huang et al. 2017).

Therefore, the land finance policy and land expropriation system interact with the decisions and actions of migrant workers to produce new structural properties, which cause the exclusion of migrant workers from accessing main transit networks, fast transport modes and distant social activities during the process of suburbanisation.

6.8 The Suburbanisation Policies Facilitate the Co-location Between Industrial Jobs and Migrants' Housing

District governments have played an important role in the suburbanisation of Shanghai. Government senior officials are evaluated by their promotion of industrial suburbanisation, foreign direct investments, industrial production and GDP growth through the establishment of various industrial clusters, new towns and other function zones (Tian et al. 2017a, b, c). To tap into cheap labour sources for industrial production, district governments always build affordable housing near industrial parks for migrant workers. For example, the Songjiang District government engages in business partnerships with Foxconn Technology Group, the world's largest electronics manufacturer, to build factories in the Songjiang District by providing the company with access to cheap land, bank loans, and affordable housing for the workers.

The author visited the Foxconn factories in Songjiang. Figure 6.4 depicts Foxconn's industrial compound. The Songjiang sub-district government built affordable housing for Foxconn's migrant workers along Nanle Road. In Songjiang, the Foxconn Technology Group provided several thousand jobs for migrant workers, though some did not want to live in dormitories because they wanted to live with their family members and take care of their children. Thus, the government policies of build housing and factories interact with the actions migrants take to live near manufacturing jobs to produce the jobs-housing co-location effects process. This jobs-housing co-location process also shows the cycle of structure and agency that contributes to the social exclusion of the migrants. Migrant workers limit their social and recreation activities to the migrant enclaves. They do not have adequate time intervals to establish social contact with the local residents in the host society.

Fig. 6.4 Foxconn large-scale factory and the nearby affordable housing for migrant workers. Source: The author took the photographs

6.9 Migrant Enclave Development in the Inner Suburbs

In 1990, downtown Shanghai served as the major receiving centre for rural migrant workers; approximately 510,000 migrants lived in the city, seeking higher income, and they accounted for approximately 3.8% of the total permanent population (Shanghai Statistics Bureau 2017a). Some of the migrants felt unable to afford the rents in shantytowns, so they slept on the sides of the roads. The *hukou* system prevented the migrants from buying houses, finding formal jobs and accessing social services (Liao and Wong 2015; Wu 2008).

The actions of migrants modified the suburbanisation policies through the migrants' reproduction of routine practices. When the industrial jobs moved to the suburbs, the migrants followed the jobs (Fig. 6.4). Some of them lived in the dormitories provided by factories (Liao and Wong 2015), while others lived in the apartments built by the villagers and found jobs in TVEs. Urban villages were dispersed within the inner suburbs.

In addition, social networks are the structures (e.g., rules and resources) that contributed to the concentration of migrant workers in the inner suburbs. A migrant's first place of residence in Shanghai is largely predetermined by the location of his or her kin or friends. Migrants rely heavily on their social relationships to secure job opportunities. Therefore, the interactions between social networks, industrial suburbanisation and the actions of migrants produce new structural properties, including the development of a system to find jobs, help each other and fight against social discrimination. One of the bottom-up emergences of the migrants is the formation of migrant enclaves in the inner suburbs (Fig. 6.5). Migrant enclaves

Fig. 6.5 Enclaves developed in the form of small shopping complexes and stores in the Minhang and Songjiang districts. Source: The author took the photographs

refer to a geographical concentration of migrant workers with the same hometown origin. Migrant enclaves are generally developed in the form of self-employed food retail businesses, such as those in the small shopping complexes and stores in the Minhang and Songjiang districts, where a shopping complex comprises shops and residential flats. Shops are built on the ground level, and residential units are built on the upper floors. The type of businesses in the shopping complex varies greatly, from grocery shops to restaurants and bakeries (Fig. 6.5). The development of enclaves contributes to migrants' short distance travel behaviour because they live and work in the same neighbourhood. The process of enclave development is a self-organisation process that produces new travel behaviours that contribute to the declining ridership of rail transit services in the suburbs.

6.9.1 *Transformation of Enclaves into New Employment and Retail Streets*

Some suburban districts have evolved into new regional employment sub-centres, where most migrants are concentrated around factories and food retail centres. Some enclaves have evolved into wholesale walking streets, such as Songjiang Zhongshan Road and prosperous shopping streets near the Zuibaichi Park Rail Station. The development of the prosperous retail and shopping centres in the suburban districts indicates that migrants' incomes have increased and they can afford to consume food and clothing and participate in leisure activities. Some migrants borrow start-up cash from their relatives to develop small retail business in the enclaves. The emergence of retail streets and employment centres in the inner suburbs is the product of the self-organisation process that includes, for example, the interaction between the policy to demolish urban villages and the actions of migrants. Many migrants switch to living in private flats in the enclaves and find retail jobs after their homes in urban villages are demolished. They either seek jobs in the retail centres or borrow start-up cash for developing small retail businesses near railway stations.

6.10 Demolishing Urban Villages to Curb Urban Sprawl

In the face of urban sprawl, the central government developed a land quota system for the Shanghai Government to follow so that the municipal government could create extra land quotas to be used to secure a supply of land in major settlements. The land quota system allows the land quotas for construction to be interchangeable between rural and urban areas. Accordingly, if a certain amount of construction land in suburban areas is reclaimed for agricultural use, an equal amount can be added to the total quota for construction in urban areas. This policy was originally designed to strengthen the preservation of agricultural land, but it has driven local governments to claim farmers' housing plots and TVEs in rural areas for agricultural use, as the creation of new agricultural land allows the district governments to sell an equivalent amount of land (Tian et al. 2017a, b, c).

For example, Dongmencun was an urban village located on the eastern side of the Zhubaichi rail station in the Songjiang District (Fig. 6.1). The village was demolished because of the land quota system. The villagers in Dongmencun were resettled and allowed to acquire two or three resettlement housing units in the relocated area as compensation; they usually lived in one housing unit and rented out the others to migrant workers. The author visited Dongmencun in 2014; it was an urban village with self-employed business owners and many rural migrant residents, who mostly found manufacturing jobs in the nearby Songjiang ETDZ. When the author visited the village again in 2018, it had been demolished and all the residents had left (Fig. 6.6). The policy of the demolition of inefficient rural construction land, such as urban villages and TVEs, results in many migrant workers in Shanghai losing their homes and their means of subsistence. After Dongmencun was demolished, its migrant residents were displaced even farther away, making it less likely that they could find a job similar to the one they had previously. Thus, the policy interacts with the actions of migrants to produce retail and manufacturing employment centres, such as the shopping streets near the Zhubaichi rail station (see Fig. 6.12).

Fig. 6.6 Dongmencun (Songjiang Dist.) in 2014 before demolition and in 2018 after demolition. Source: The author took the photographs

6.10.1 Migrants Live in Low-Cost Private Housing

The results of surveys on rural migrants in the Jiading and Songjiang districts indicated that approximately 76% of rural migrant households rented private housing, and over 80% of them were unsatisfied with their crowded and small living environment (Jiading District Statistics Bureau 2014; Songjiang District Statistics Bureau 2014). Based on survey data of Shanghai's migrant population in 2011, approximately 88% of this group earned less than 2500 RMB per month, or approximately 57% of the local working residents' monthly income (Yang et al. 2015; Shanghai Statistics Bureau 2012). These migrants their choice of residence, they are restricted to low-cost housing.

In response to the migrants' increasing demand for private housing, a great deal of newly developed commodity and vacant resettlement housing has also entered the rental sector. Studies have found that many high-rise small-unit apartments are built

Fig. 6.7 Migrants' housing near Zuibaichi park station, Songjiang district. Source: The author took the photograph

and clustered along metro lines to attract those employed in the city centre. In addition, many apartments that are bought as property investments have been rented out, fitting perfectly with migrants' high demand for housing. An apartment with two or three bedrooms can be sub-divided and separately leased to several migrant families. This is called co-renting and has become prevalent in the suburbs, such as in the apartments in the Zhubaichi neighbourhood (Shen and Wu 2013; Shen 2011) (Fig. 6.7).

6.11 Rural Migrants Take Short-Distance Trips

The interaction between policies and migrants' practices produces new structural properties, such as living in enclaves, finding industrial and retail jobs, living near employment and taking short-distance trips. Most of the migrants' daily activity are spatially restricted to the enclaves. For example, in 2014, the results of a survey on rural migrant workers in the Songjiang District found that among the 205 respondents, 75% spent approximately 10 min commuting to work, and only 6% spent more than 30 min on work trips. Approximately 80% of the migrant respondents either rode electric bikes or walked to work (Songjiang District Statistics Bureau 2014). In addition, a similar survey in the Jiading District indicated that most migrant respondents chose to live near their jobs, and their average travel time for work trips was approximately 15 min. Approximately 48.5% of the respondents did not own Shanghai public transportation auto-payment cards, which indicates that they seldom used buses or rail transit services for long-distance commuting (Jiading District Statistics Bureau 2014). According to a survey in five large-scale residential areas in the suburbs, approximately only 20% of the migrant respondents used rail transit services to reach work (Guan and Xu 2018).

The self-organisation process involving the interaction between structures and the agency of low-income workers produces new structural properties, which include the suburbanisation of poverty. The structures include the *hukou* system, urban redevelopment and the property-led redevelopment approach, the suburbanisation of industries, the development of a rail transit system and the land finance policy. In addition, the social networks and norms of migrants contribute to enabling migrants to find jobs and develop enclaves near industrial and retail employment centres. These structures interact with the migrants' decisions and actions, resulting in them following employment and suburbanising the inner suburbs, depending on social networks to find jobs and concentrating in enclaves to avoid social discrimination. Because of the self-organisation process, migrant workers not only are excluded from the mainstream activities of society but also experience accessibility problems in reaching non-work activities. Furthermore, the migrants' short distance travel behaviours give rise to the low level of ridership that is insufficient to support cost-effective rail transit services in the suburban areas of Shanghai.

6.12 Transit-Oriented Development in Shanghai Is a Product of the Self-Organisation Process

Suburbanisation is regarded as a self-organisation process in which structural factors, such as the rail transit system, interact with the actions of relocated residents and migrant workers to produce new structural properties. The new properties include the land sales revenue being used to build rail transit networks to transport travellers between the urban centre and new towns. Other new structural properties include the networks of the rail transit system (the Shanghai Metro) being unable to cover all of the housing and job locations of the displaced residents and migrants. Many relocated local residents and migrants are excluded from accessing fast transit systems, resulting in a reduction in trip rates and travel distances to fulfil basic needs. On the other hand, the high level of mobility provided by the Shanghai Metro encourages many relocated migrants, investors and local residents to relocate to new towns. The actions of these actors facilitate the development of employment centres in the suburbs. New towns are usually developed using mixed land-use development, with medium densities of commercial buildings and residential resettlements around major rail transit networks. The land-use development and industrial zones also generate potential population growth and thus employment centres for migrant workers. The self-organisation process also produces new structural properties, which include the increase in land prices near rail transit lines. The district governments then use the new town development to leverage bank loans. The developed land that has high accessibility is put up for sale through competitive bidding. However, this land-use and transport integration in new town development faces problems because the routine travel demand for long-distance trips between new towns and the city centre is not sufficient to support cost-effective rail transit services (Pan et al. 2011).

Based on the concept of the self-organisation process, this chapter attempts to investigate the root causes of insufficient rail ridership. In the suburbs, the activity centres, that surround the rail transit stations, such as offices and businesses, are relatively less dense than those in the downtown areas. This indicates that the suburban areas fall short in terms of employment in services and businesses (Pan 2009). The transport literature has indicated that the performance of rail transit services depends on land-use development around station areas, which is developed on the same level as the TOD design guidelines. The density of employment, residential and retail resettlement around the rail station areas determines the rail travel demand, which is an important factor for cost-effective rail transit services (Cervero et al. 2013, 2017).

Many relocated local residents have to change from non-motorised to transit commuting in order to reach their jobs located in the city centre. In addition, longer commuting distances are strongly associated with a higher probability of riding rail transit services to work (Shen et al. 2016; Cervero and Day 2008). Therefore, the main purpose of building rail transit networks is to provide rapid and comfortable rail transit systems for relocated working residents so that they can maintain strong commuting connections between their homes in suburban areas and their employment in downtown areas (Shen et al. 2016; Pan et al. 2011).

6.13 Route Test on the Shanghai Metro

This section introduces a route test that was based on an isochronal map approach to determine the mobility provisions of the Shanghai Metro for low-income residents in suburban areas. The isochronal map is formed by isochrones. An isochrone ('iso' means equal, and 'chrone' means time) is a line drawn on a map that connects the locations to which travellers commute by taking rail transit services from a common origin (e.g., a point in the city centre) to the locations in the city or the suburban areas that have the same travel time. This study selected the People's Square or the Bund in Shanghai's city centre as the focal point of the route test (Fig. 6.8).

The route-test survey of the Shanghai Metro was conducted in January 2019. The information on travel times was based on Google Maps (Google Maps 2019) and the information on the fare tables provided by the Shanghai Metro (Shanghai Metro 2019). The results indicated that travellers living in Shanghai's urban districts, i.e., Huangpu, Changning, Jing'an, Zhabei, Hongkou and Xuhui, could reach their workplaces in the Bund neighbourhood within 40 min using the Shanghai Metro

Fig. 6.8 Isochronal map of taking the Shanghai Metro. Source: Modified from Google maps (2019)

(Fig. 6.8). The fare for a single trip on the Shanghai Metro within these urban districts was approximately RMB 5 (US$0.73). Thus, the Shanghai Metro provided a high level of mobility for travellers in urban districts. Transport costs on rail transit services should be affordable for low-income workers (Fig. 6.8).

Furthermore, travellers living in the inner suburbs, such as the area surrounding Hongqiao Airport, Putuo, Gucun and large parts of Minhang and Jiading, were able to reach their jobs in the Bund neighbourhood within an hour (Fig. 6.8). In 2016, of five categories of workers in Shanghai, low-income workers received the smallest disposable income (an average of RMB 2017 or US$296 per month). The fare for a single trip on the Shanghai Metro within this travel time zone was approximately RMB 5 (US$0.73). It is predicted that the travellers in the inner suburbs had to take feeding bus services to the rail stations and that the fare of a single feeding bus trip was RMB 2 (US$0.3). Thus, the low-income workers spent approximately 13.9% of

Fig. 6.9 Travellers park their E-Bikes outside the Songjiang sports centre station. Source: The author took the photograph

their monthly disposable income on daily work trips between the city centre and the inner suburbs (Shanghai Statistics Bureau 2017a). Many low-income workers had been displaced to the resettlement public housing in this commuting time zone, such as Pujiangzhen in the Minhang District, Jiangqiao in the Jiading District, and Sijing in the Songjiang District (Guan and Xu 2018) (Fig. 6.8).

In Shanghai, a hub-and-spoke network framework was introduced for the multi-modal transit system. Spoke or feeder routes were used to connect the suburban neighbourhoods and rail transit stations. The modes of transport used in these feeder services are public buses and taxis. According to the author's observations, for residents in the suburbs, the feeding bus services for the Shanghai Metro were extremely inadequate. Instead, many travellers rode-bikes or drove private cars to the rail stations, parking their cars near the rail stations and taking the trains to the city centre (Fig. 6.9).

Travellers living in suburban areas, such as the new towns in the Baoshan District and the Minhang District, Xinqiaozhen in the Songjiang District, and the area surrounding Pudong Airport, were able to reach the Bund within 1.5 h. The low-income workers living in Shanghai's main inner suburbs, such as Antingzhen, could reach their workplaces in the central business district within 2 h. The fare for a single trip on the Shanghai Metro within this travel time zone was approximately RMB 9 (US$1.3), and the low-income workers spent approximately 21.8% of their monthly disposable income on daily work trips between the city centre and their homes in the inner suburbs, including the fares for feeding bus services (Shanghai

Statistics Bureau 2017a). The relocated low-income local residents who live in this travel time zone likely feel that the transport costs are unaffordable. In addition, if they want to find jobs in the city centre, they must spend 2–3 h each day on the rail transit services to travel between their homes and jobs. Therefore, they not only experience accessibility problems in accessing employment but are also excluded from accessing social networks and community connections that are mostly located in downtown.

The travellers who live in distant new towns that are not covered by the Shanghai Metro network, such as Fengxian and Luchaogangzhen, have to transfer repeatedly between buses and the Shanghai Metro, spending approximately 4 h to reach the Bund (Fig. 6.8). The results of the route-test survey indicated that the Shanghai Metro provided adequate mobility for the low-income workers who lived near the rail stations in the inner suburbs or lived in the city centre. However, for those who lived farther away from the rail transit networks likely faced mobility problems. These workers likely feel that the rail service fares are unaffordable and the frequency of feeding bus services is inadequate (see also Yuan and Lin 2013; Day and Cervero 2010).

The results of the route-test survey confirm that rail services cannot resolve the jobs-housing spatial mismatch accessibility problems that are caused by the suburbanisation process. The suburbanisation process produces new structural properties, which include the sprawling housing locations and inadequacy of the Shanghai Metro's network density to cover all of the necessary locations. The relocated local residents use their individual decisions and actions to transform the rail transit services. Many displaced residents with Shanghai *hukou* status find jobs near their neighbourhoods of residence to save on transport costs (Day and Cervero 2010). Many of the older displaced residents choose to take early retirement and apply for social welfare to avoid commuting long distances (Guan and Xu 2018; Yuan and Lin 2013). Younger relocated local residents tend to switch from non-motorised to rail transit commuting (Cervero and Day 2008). Regarding migrant workers, Guan and Xu (2018) find that in general, most migrants in the suburbs take short-distance trips and do not choose rail transit.

6.14 The Discrepancy Between Actual and Forecasted Ridership of the Rail Transit Services

Recent transport studies have suggested that the real travel behaviours of displaced local workers and migrant workers are not in line with the estimated travel patterns suggested by conventional transport planning models, resulting in the declining ridership of rail transit services in Shanghai (Pan et al. 2011; Wang et al. 2012). A high number of the relocated residents with Shanghai *hukou* status are low-income workers who involuntarily moved to the suburbs. Most of them are in 50–60-year-old age groups and do not work. They tend to travel by bus and e-bike to reach activities, and a large share of them do not use rail transit services. This is because their job accessibility levels declined dramatically following their move (Guan and Xu 2018; Yuan and Lin 2013; Cervero and Day 2008). Another group of relocated residents in the suburbs are migrant workers, who accounted for approximately 33.1% of the population in the suburban districts in 2016. These residents tend to take short-distance trips, and thus most of them do not take the Shanghai Metro to reach work (Guan and Xu 2018; Shanghai Statistics Bureau 2017a). The author argues that the individual decisions and actions of the relocated workers have far stronger influences on the declining rail ridership than the planning rules and the built environment, such as the TOD design guidelines around the rail station areas. The author conducted on-site surveys to support this argument, which will be discussed later in this chapter.

Migrants take short-distance trips because of the self-organisation processes that are taking place in migrant enclaves. For example, using cellular network data, one study quantified Shanghai's jobs-housing balance with a larger sample (n = 6.3 million). The study examined the commuting patterns of different subareas in Shanghai (Zhang et al. 2017). For those travellers whose workplaces were located in the suburbs, their average travel distance was 7.7 km. The short travel distances of these travellers were due to approximately 67% of this group of travellers also living in the suburbs. In addition, for the other group of travellers who lived in the suburbs and found employment in the city centre, their average travel distance was 9.0 km. The jobs-housing ratio (e.g., a ratio of jobs to dwelling units) in the suburban areas was 0.82, which indicated a high level of jobs-housing balance (Zhang et al. 2017).

It is estimated that between 20% and 26% of the travellers who lived in the suburbs and found jobs in the city centre took rail transit services to work (see also Guan and Xu 2018, Yuan and Lin 2013). The travel practices of the travellers in the suburbs directly caused the declining rail ridership that was insufficient to support cost-effective rail transit services.

6.15 Influences of the Built Environment and TOD on Commuting in Shanghai

The built environment refers to the human-made space in which people live, work, and engage in recreational activities on a day-to-day basis, and it includes the environment that is built in accordance with TOD design guidelines. One study found that the built environment in Shanghai has a strong impact on travel behaviour. The study uses the data collected from 2038 residents currently living in TOD and non-TOD neighbourhoods in Shanghai, finding that the high density of commercial and residential development in the study neighbourhoods is the main factor influencing residents' travel behaviours (Chen et al. 2017). For example, residents in TOD neighbourhoods were surveyed, and the results indicate that nearly 70% of commuting trips are made by rail transit and non-motorised modes. This is nearly two times higher in comparison to the transport modes of non-TOD residents (Chen et al. 2017). Therefore, policies that use TOD design guidelines as the basis for developing land use will increase the ridership on rail transit services (Cervero et al. 2013; Wang et al. 2012).

A study carried out by Wang et al. (2012) in Shanghai found that by estimating the railway traffic demand produced by planned land uses, the railway transport plan could, in general, meet the transport needs of suburban communities in Shanghai, although mismatch still existed in some areas. Pan et al. (2017) found that insufficient rail ridership emerged in the suburbs and was mainly due to the land-use development of the railway stations not developing on the same level as the TOD design guidelines. Pan et al. suggested that commercial development should be integrated into the planning of rail transit stations in the inner suburbs. In addition, there were no adequate commercial activity centres around the rail transit stations in the suburbs.

The author of this book argues that Shanghai transport planners should have followed the design guidelines of TOD when they planned and built the Shanghai Metro. Based on the TOD concepts, the medium intensity of residential and retail development was developed around rail station areas in the suburbs. These studies also found that most working residents in the suburbs tended totravel short distances as part of their work obligations and they did not take rail transit services to work (Guan and Xu 2018; Pan et al. 2011).

Based on the TOD design guidelines, this section describes the on-site surveys conducted and tries to evaluate whether the land-use development of the rail stations developed on the same level as the TOD design guidelines. This book argues that if the land-use development around the rail stations in the suburbs is matched by the TOD design guidelines, then the built environment around the rail stations is not the main cause of the declining ridership of the rail transit services. The TOD guidelines include eight key objectives (Cervero et al. 2017, p. 121):

- Walking: a public realm that is safe, complete, active, comfortable, and vibrant;
- Cycling: a cycling network that is safe and complete, with ample and secure parking;
- Connectivity: walking and cycling routes that are short, direct, and varied, more so than driving routes;
- Transit service: high-quality transit that is accessible by foot;
- Mixed land uses: diverse and complementary land uses that shorten trip lengths;
- Density: residential and job densities sufficient to support high-quality transit;
- Compactness: developments that infill built-up areas and improve access to other transit interchanges; and
- Shifting: land devoted to off- and on-street parking and driveways is reduced.

6.15.1 On-Site Surveys to Evaluate Whether the Land-Use Development around Rail Stations Is Matched by the TOD Design Guidelines

Pan et al. (2017) find that rail stations, which are better integrated with nearby commercial development, tend to have larger passenger volumes. In addition, passenger volume is positively associated with employment density around rail stations. Wang et al. (2012) suggested that by estimating the railway traffic demand produced by planned land uses, the TOD plan can, in general, meet the transport needs of suburban communities in Shanghai. Based on the above study findings, this section tries to determine whether the land-use development around rail stations is developed on the same level as the TOD design guidelines.

Based on the investigation approach proposed by Schuetz et al. (2018), three on-site surveys were chosen for the on-site survey activity. In the on-site surveys, the author documented the densities of the population and resettlements near stations, as well as the amount and type of business and retail development around the rail stations. The results of the surveys illustrate that the rail station land-use development has followed the TOD guidelines and that the related land-use policies near the study rail stations could spur ridership on the rail services. The author argues that TOD-friendly development alone provides insufficient ridership for supporting cost-effective rail services in suburban areas. The social exclusion processes in the suburbanisation of Shanghai that produced the short-distance travel behaviours of the affected low-income workers are the root cause of the declining ridership of the rail transit services.

The author investigated residential and employment densities, vehicle parking, sidewalk provision, bicycle facilities, and pedestrian connectivity around the three study rail stations. The three study rail stations were Anting Station (Line 11) in the Jiading District, Lianhua Road Rail Station (Line 1) in the Minhang District and Guibaichi Park Station (Line 9) in the Songjiang District.

6.15.2 Anting Rail Transit Station

Figure 6.10 depicts Anting Rail Station (Line 11), which is located in the Shanghai Automobile City of Anting New Town in the Jiading District. The Shanghai Master Plan (1999–2020) selected the town of Anting for developing the automobile industry. Shanghai Metro Line 11 serves Anting (Fig. 6.1). On the side of the station, there are large-scale automotive plants, such as the Shanghai Volkswagen Automotive Plant. The plants employ more than ten thousand rural migrant workers. Most of the workers in these auto-manufacturing plants either live in the company-provided dormitories or live in the nearby low-income neighbourhoods built by the sub-district government. There are some parking facilities at the plant entrances, and many electric bikes are parked near the factory entrances. As the author walked

Fig. 6.10 Shanghai Volkswagen automotive plant near Anting Rail Station (Line 11) in the Jiading district. Source: The author took the photographs

along the main road, many construction migrant workers were constructing high-rise commodity housing near Anting Rail Station. They had their lunch near the construction sites while the author was investigating their behaviours. Some temporary buildings for the migrant workers had been built. The local government built resettlement structures for construction workers and their family members, all of whom were migrants.

The land-use development around the rail station areas showed that the rail transit system improved the accessibility of the neighbourhood around Anting Rail Station, resulting in higher land prices and the attraction of a higher density of residential and industrial development, such as new high-rise buildings and auto-making factories. There were some retail shops, restaurants and hotels. Li et al. (2019) find that unit prices for apartments within the 0.5-mile buffer zone of metro stations were significantly higher than the apartments farther away from metro stations, and this relationship was more significant for apartments located in the suburbs. Thus, the land price of the housing around the Anting Rail Station should be higher than the housing in other areas in the suburbs, thus encouraging investors to develop a medium density of residential development near the station areas.

A bus interchange was developed near the Anting Rail Station (Fig. 6.10). The bus routes were used to transport workers between the automotive plants, rail station and homes of the workers. Some of the travellers took the transit busses for shopping and recreational purposes. The author boarded a bus that transported him to nearby affordable housing neighbourhoods, such as the Huaqiao neighbourhood for migrant workers and the Xi'an community where many three-storey old houses were built on both sides of the road. Many of these private houses were rented out to rural migrants. The ground level of the houses was used for retail shops and manufacturing production. The upper levels of the houses were used for residential purposes. Eight to ten blocks of high-rise buildings (every 20–40 stories in height) were built approximately one mile away from Anting Rail Station. The high-rise buildings were most likely public housing for the relocated local residents who moved from urban areas to the suburbs during the process of urban redevelopment. In addition, several high-rise hotels, shopping malls and office buildings were developed near Anting Rail Station and were built for foreign investors, visitors and retail business activities.

The author argues that the high and medium residential and retail settlement densities around Anting Rail Station would have provided sufficient travel demand for rail services to support cost-effective services if most of the residents in Anting had sought employment in the city centre and used the rail transit services. However, migrant workers accounted for approximately 57.4% of the total population in the Jiading District (Fig. 6.2), and most of them chose to find jobs in auto manufacturing, retail and construction activities near their homes due to the interactions between the *hukou* system and individual actions of the migrants.

Regarding relocated workers with Shanghai *hukou* statuses, many of the young residents took rail services to travel from the relocation housing estates to their jobs in the city centre. Some middle-aged local residents who had been displaced could not find new jobs in Anting, and they chose to take early retirement. The self-organisation process produced new structural properties, which included the migrants travelling shorter distances for work trips and finding jobs near their homes. The older relocated workers with Shanghai *hukou* status used slow transport modes to reach nearby recreational and medical activities. The new travel behaviours of the migrant workers and displaced local workers in Anting contributed to the low level of ridership that is insufficient to support cost-effective rail transit services in the suburbs.

6.15.3 Lianhua Road Rail Transit Station

Figure 6.11 depicts Lianhua Road Rail Station (Line 1), which is located in the Minhang District (Fig. 6.1). Near the station, there is a large park-and-ride facility for passengers to park their electric bikes and take rail transit services. The author

Fig. 6.11 Lianhua Road Rail Station (Line 1) and Bus interchange in the Minhang district. Source: The author took the photographs

observed that many electric bikes were parked in the parking spaces. On the opposite side of the station, there is a shopping mall with a large parking lot for private cars. In addition, several commercial buildings (each 15–20 storeys in height), including a hotel, were built around the station. Furthermore, a bus interchange was developed near the station, and it provided free feeding-bus services for local residents to take rail services; rail passengers can take several bus routes to the nearby housing neighbourhoods (Fig. 6.11).

There are sufficient electric bike lanes and pedestrian walkways to connect the station with other nearby residential neighbourhoods. In the north-eastern part of the station, several middle-income housing neighbourhoods have been developed; most of the houses are between three and ten storeys in height. Several houses are grouped together in the form of gated communities to enhance the local residents' security. Within the neighbourhood, the residents use several main transport modes to reach activities; they drive private cars, take buses, ride electric bikes and walk. There are road networks for private cars and buses and off-road lanes for electric bikes and pedestrians using the sidewalks. Many migrant workers ride electric bikes to the station, and only a few pedestrians walk on the streets. A few migrant enclaves have been developed in the form of small, three-storey shopping buildings with food sellers, private car repair shops and other retail businesses. The shopping buildings are located among the gated housing communities to provide retail services for the residents. On the western side of the station, there are several blocks of tall commodity housing, some of which are rented out to migrants.

Most of the buildings around the station are used for residential purposes. The commercial buildings on the opposite side of the station provide many retail and business job opportunities that generate passenger demand for the rail services in Lianhua Road Rail Station. The tall commercial buildings, together with the low-rise middle-income houses, have produced compact and mixed land-use development around the station areas. The author argues that the land-use development would have provided sufficient travel demand for cost-effective rail transit services if most residents had sought employment in the downtown areas and had used the rail transit services. However, migrant workers account for approximately 50% of the total population in the Minhang District (Fig. 6.2), and most of them choose to find retail and manufacturing jobs near their homes due to the interactions between the *hukou* system policies and the individual decisions and actions of the rural migrants. The better-off local residents tend to drive private cars to work; thus, only the young and low-income local residents take rail services to reach jobs in the urban centre. Most of the local residents and migrants generate short-distance work trips, and their travel practices contribute to the low level of ridership that is insufficient to support cost-effective rail transit services.

6.15.4 Zuibaichi Park Rail Station

Figure 6.12 depicts Guibaichi Rail Station, which is one of the stations along Shanghai Metro Line 9. Guibaichi Rail Station is located in the Songjiang District (Fig. 6.1). Several feeding bus routes provide linking services for the rail transit station, and the author observed that there were many passengers waiting for buses at several bus stops. Near the station, there are park-and-ride facilities for passengers to park their electric bikes and take the rail transit services. There is sufficient sidewalk provision for pedestrians. The neighbourhood that surrounds the station areas has adopted mixed land-use development, and most of the residents are industrial and retail workers. The industrial jobs are provided by the Songjiang ETDZ, which is located approximately 2–3 km from the station and includes the large-scale Foxconn industrial compounds.

High-density retail establishments were developed approximately 500 m away from the station (Fig. 6.12). The shopping street is characterised by a high density of local retail businesses, including restaurants, a medium-scale hotel, supermarkets, boutiques and fast-food chain services (e.g., McDonald's and Kentucky Fried Chicken). The retail employment in the shopping street is driven by sufficient

Fig. 6.12 Retail establishments near Zuibaichi Park Rail Station (Line 9) in the Songjiang district. Source: The author took the photographs

affordability and the increasing income of customers, who are mostly rural migrant workers (Fig. 6.12).

The high-density retail settlements indicate that Zuibaichi Park Station has increased the accessibility of the surrounding neighbourhoods, leading to high land values for residential and commercial development. Li et al. (2019) find that proximity to railway stations in Shanghai has a positive effect on land prices (Fig. 6.13). In addition, commercial properties near rail stations have higher land prices than residential properties. Zuibaichi Park Station is located in rural migrant neighbourhoods, and thus, the land-use development around the station should be transformed from migrant enclaves into busy shopping streets by the actions of migrant workers and investors. The high accessibility of the neighbourhood has also attracted middle-class families who previously lived in nearby gated communities, such as the middle-class housing neighbourhoods near the Songjiang New City and Songjiang Sports Centre Rail Stations. The residents in these neighbourhoods seek

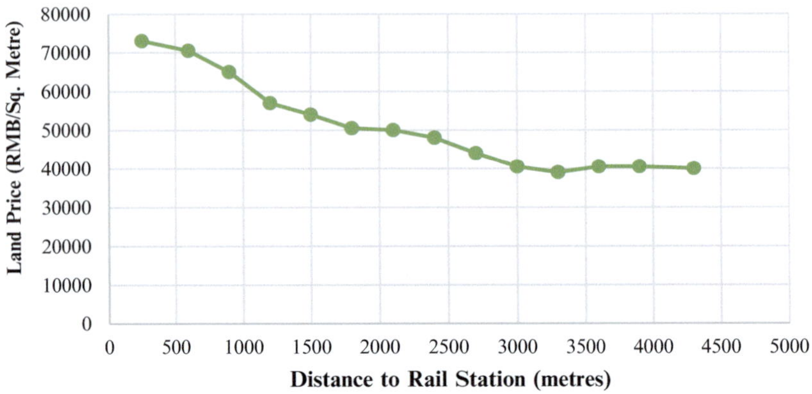

Fig. 6.13 Rail transit station proximity has a positive effect on property value in Shanghai. Source: Modified from Li et al. (2019)

out recreational and shopping activities by taking rail services to Zuibaichi, and they have contributed to turning the neighbourhood into a shopping centre.

As the author walked approximately 1 km from the station, there were a few high-rise buildings that were up to 20 storeys tall and many low-rise buildings surrounding Guibaichi Rail Station. The author argues that the high and medium residential and retail settlement densities would have provided sufficient ridership to support cost-effective rail transit services if most of the residents had sought employment in the downtown areas and had used the rail transit services. However, migrant workers account for approximately 61.3% of the total population in the Songjiang District (Fig. 6.2), and most of them choose to find retail and manufacturing jobs near their homes due to the interactions between the *hukou* system and the migrants' individual actions. Thus, the migrants and most local residents in the Songjiang District tend to travel short distance trips to reach activities. The interactions between the agencies of the relocated residents and the government policies produce new travel behaviours that have resulted in the declining level of ridership that is insufficient to support cost-effective rail transit services. In fact, new travel behaviours are produced by the suburbanisation of poverty, which is also a self-organisation process.

6.16 Summary of the Research Results

The ridership of rail transit services is shaped by the social exclusion process, which involves a multi-dimensional phenomenon (Stanely et al. 2019; Schwanen et al. 2015). The government policies on the suburbanisation of Shanghai have influenced the commuting of relocated local workers and migrants. The policies interact with the individual decisions and actions of local workers and migrants to produce new structural properties, which include new travel behaviours that negatively influence

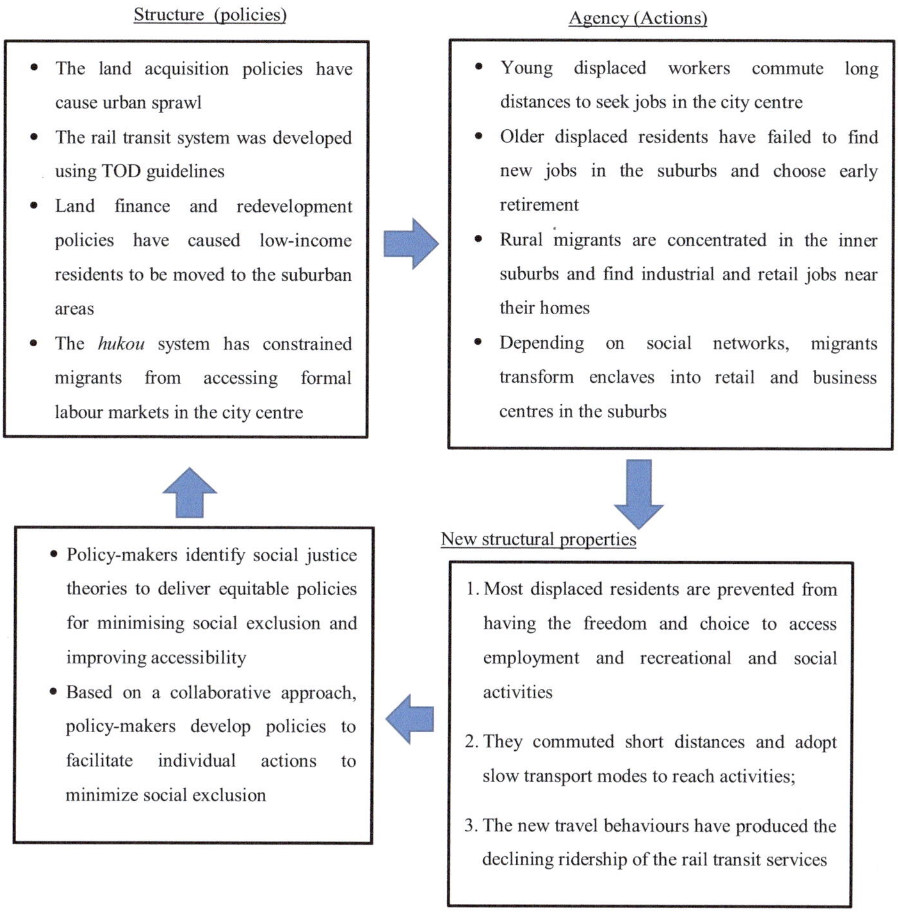

Fig. 6.14 Suburbanisation of poverty process produces new behaviours in Shanghai

the rail ridership and cause problems in achieving cost-effective rail services in the suburbs.

In addition, the social exclusion process can be investigated in the form of a two-stage process involving top-down emergence and bottom-up emergence (Fuchs 2003). Top-down emergences, such as land financing, the *hukou* system, urban redevelopment and the acquisition of farmland policies, has caused urban sprawl and accessibility problems. Another top-down emergence is the provision of rail transit services (the Shanghai Metro) to connect urban areas and new towns. The rail transit networks have failed to cover many employment and housing locations because of the urban sprawl that resulted from the process of suburbanisation (Zhang et al. 2018; Tian et al. 2017a, b, c; You 2016). Many low-income displaced workers who want to take the rail transit services to the city centre have to take feeding bus services, pay unaffordable transit fares, and experience long travel times (Fig. 6.14).

The property-led urban redevelopment policy is a top-down emergence that has relocated low-income workers from the inner city to the suburbs. Many low-income relocated residents with Shanghai *hukou* status are involuntarily movers; they live in resettlement housing that is far away from their workplaces and mostly located in the downtown areas, resulting in commuting problems. They also live far away from the social networks that they had developed in urban areas (Yuan and Lin 2013; Cervero and Day 2008; He and Wu 2005, 2007). The redevelopment policy thus constrains the displaced local workers from having the freedom, choice and opportunity to reach social activities and jobs. Both migrants and displaced workers are constrained in exchanging their labour power for adequate income and food (Fig. 6.14).

In the face of these transport-related social exclusion problems, many relocated residents and migrants have developed bottom-up emergence. They choose to find jobs near their homes and avoid taking rail transit services to reach distant employment opportunities. In terms of jobs-housing spatial mismatch accessibility problems, no more than 30% of the displaced local residents commute long distances to the city centre (Guan and Xu 2018; Yuan and Lin 2013). Many of them have reached retirement age, so they have chosen to take early retirement and avoid commuting long distances to seek distant employment in the city centre. The daily commuting purposes of this group of relocated local residents tends to be concentrated on non-work activities such as shopping and recreational and social activities. They take slow transport modes such as buses and e-bikes (Fig. 6.14) (Guan and Xu 2018; Yuan and Lin 2013).

Because of the policies on collective-owned land acquisition and the sprawling of industrial zones, most of the migrants' housing neighbourhoods are not located near rail transit networks. The migrant workers have developed enclaves in the inner suburbs to avoid social discrimination, and they depend on social networks to find jobs. Some of them have started self-employed retail shops. Therefore, most of them live and work in neighbourhood enclaves. They also take short-distance trips and choose slow transport modes to reach social and recreational activities (Fig. 6.14).

In addition, some young relocated local residents and migrants who live near rail stations in the suburbs are not influenced urban sprawl problems, and the rail transit system enables the commuting of these young travellers from their homes in the suburbs and their jobs in the city centre (Cervero and Day 2008). The interactions between government policies and the agency of low-income workers produce the new structural properties and travel behaviours that cause the low level of travel demand that is insufficient to support cost-effective rail transit services (Fig. 6.14) (Guan and Xu 2018; Yuan and Lin 2013).

Furthermore, the built environment is a top-down emergence and contributes to shaping the performance of rail transit services. The on-site investigations in this chapter on the three rail stations indicated that the built environment around the stations provides a certain amount of travel demand for the rail transit services because the built environment around the three rail stations is in line with TOD design guidelines. In regard to the bottom-up emergence of the travellers, many working residents who lived near rail stations use rail transit services to commute to employment opportunities and shopping and leisure centres. Travellers ride and park their e-bikes near the rail stations. Some migrants and local residents even invest in and develop shopping and retail streets around the rail station areas. However, the results of the site surveys also indicated that the interaction between the built environment and the agency of displaced residents alone cannot provide sufficient ridership to support cost-effective rail services in the suburban areas (Fig. 6.14). The social exclusion process that led to the suburbanisation of poverty in Shanghai is the root cause of the declining ridership of the rail transit services.

6.17 Conclusion

The investigations in Chap. 6 confirm the third hypothesis of this book: the process of the suburbanisation of poverty in Shanghai has produced new travel behaviours, social exclusion, accessibility problems and the declining ridership of rail services. This chapter demonstrated a new approach to urban transport planning to investigate commuting behaviour and social exclusion. The suburbanisation of Shanghai has been a form of the transport-related social exclusion process, which is also a type of self-organisation process, as both processes involve structural factors that interact with agency to produce new behaviours. Social exclusion is a multi-dimensional phenomenon, and it shaped the investigation in Chap. 6 on the suburbanisation of Shanghai.

Poverty is an important dimension of exclusion. Government suburbanisation policies in Shanghai (e.g., the land finance policy, urban redevelopment policy and *hukou* system) have led to an overarching trend of the suburbanisation of poverty towards the inner suburbs and surrounding regions. For example, the route test on the mobility provision of the Shanghai Metro found that because of urban sprawl, the rail networks fail to cover many of the dispersed housing estates in the suburbs. Thus, the Shanghai Metro cannot provide adequate mobility to the relocated residents to commute between the city centre and the suburban areas.

The residence experience long travel times for their work trips between the suburbs and the city centre, inadequate feeding bus services, and unaffordable public transit fares. As a result, many low-income displaced workers cannot find new jobs in the suburbs and choose early retirement. They also experience exclusion from accessing distant employment and medical, recreational and social activities in the city centre. Furthermore, because of the *hukou* system, many migrants cannot find jobs in the city centre and move the inner suburbs to find informal jobs. They earn low income and have to work long hours in order to earn adequate income to support their quality of life. They generate a lower trip frequency than the general working population. They lack adequate time intervals for commuting to distant recreational and social activities.

The deprivation of personal capability is the second dimension that has caused social exclusion of low-income workers. For example, land finance and property-led urban redevelopment policies interact with the actions of the displaced low-income local workers to produce new structural properties, including many affected residents who were displaced to the suburbs living far away from their original workplaces and social networks. These policies constrain the relocated low-income workers from having the freedom, opportunity, and choice to select desirable housing locations, jobs and transport modes and exclude them from participating in the mainstream activities of society.

The other dimension is social inequities that cause the exclusion of low-income workers. The migrant workers face accessibility inequity. Industrial suburbanisation and land acquisition policies have interacted with the individual actions of migrant workers to produce the sprawling of industrial jobs and migrant homes in the inner suburbs. The process has caused the exclusion of migrant workers from accessing the non-work facilities that are near the new town centres. The workers face accessibility inequity in reaching distant locations that include employment and social activities.

In addition, research results have found that land finance and urban redevelopment policies have caused the relocation of low-income workers to dispersed housing estates in the suburbs, where they face accessibility problems in reaching work and non-work activities. Hence, these unjust policies cause the concentration of poverty in suburban areas. The on-site surveys of the three rail stations found that the social exclusion processes in the suburbanisation that the produced short-distance travel behaviours of the affected low-income workers was the root cause of the declining demand for the rail transit services.

The mechanism of the social exclusion process is the duality of structure, which means that the structural properties of social systems are both the medium and the outcome of the practices that constitute those systems (Fuchs 2003; Giddens 1981). Structure and agency are a duality; neither can exist without the other. This fundamental concept can be used to minimise social exclusion by changing policies or individual actions. The following sections will introduce strategies to change policies and individual actions to minimise social exclusion and improve accessibility in the inner suburbs of Shanghai.

The suburbanisation of poverty in Shanghai has mainly been caused by inequitable government policies. To minimise the social exclusion of low-income workers, it is necessary to evaluate the policies in terms of social justice theories. Most of the policies regarding suburbanisation in Shanghai have been based on the utilitarian approach, which allows policies to override the individual freedom of low-income workers in order to maximise productivity and the growth of the city's economy (see Pereira et al. 2017). The utilitarian approach and most of the policies regarding suburbanisation in Shanghai are incompatible with social justice.

In any attempt to determine appropriate equitable policy responses, social justice theories should be considered to minimise social exclusion (Fig. 6.14). Rawls's (1971) difference principle is proposed for use in the evaluation of policies regarding suburbanisation in Shanghai. The difference principle suggests that inequalities in the process of suburbanisation are features of every society. However, the policies regarding the suburbanisation in Shanghai will be fair only if they are to the greatest benefit of the disadvantaged, namely, the low-income displaced local workers and rural migrants (Rawls 1971).

Therefore, the municipal government should change the land-use and economic policies for suburbanisation so that relocated workers can live close to job opportunities in suburban areas, such as developing employment centres for local workers in new towns. The government should also provide dense bus and rail transit routes to cover as many relocating housing estates as possible, which will give workers reasonable mobility in commuting between their homes and their workplaces in the city centre. In addition, the government should also change the urban redevelopment policy to provide choices and opportunities for low-income residents to relocate to public housing in the city centre. As a result of that, most of the relocated workers could live close to their relatives and friends in urban areas.

According to the capability approach (Sen 1981, 1984), government decision-makers who want to build a just society in Shanghai should ensure that migrant workers in suburban areas have the freedom to achieve well-being, which includes adequate housing and employment. The author suggests that the Chinese government should change the *hukou* system so that rural migrants can find formal employment and participate in social welfare in host cities. The author argues that if the *hukou* system is changed, the self-organisation process will produce new structural properties, which could include migrant workers travelling long distances for work trips. This is because migrants will be able to find jobs and housing in different locations throughout the whole city. Furthermore, the declining ridership of rail services in the suburbs can be reversed.

In addition, the decision-makers in the Shanghai government need to understand the root causes of the individual actions of the migrants and local relocated workers in order to change the migrants' and workers' actions and attitudes to improve commuting and minimise social exclusion (Fig. 6.14). The social structure is capable of both constraining and enabling action (Giddens 1984). Healey (1997, 1998) proposes a collaborative approach for public participation to understand the individual actions and attitudes of low-income workers. As a result, policymakers can modify policies, enable action and minimise social exclusion.

Policy-makers could consult low-income workers through several informal channels, such as conducting interviews with rural migrants and relocated local workers, hosting public debates on government policies, and consulting officials from street offices and district governments about local residents' commuting patterns and job-seeking difficulties. The Shanghai government should tolerate diversity and difference, the heterogeneity of values, and lifestyle oppositions in order to understand the difficulties, skills and capabilities of relocated workers and migrants. When policy-makers understand the actions and attitudes of the affected workers, they can put the amended policies into practice to tackle social exclusion and increase the travel demand for rail transit services.

References

Alonso W (1964) Location and land use. Harvard University Press, Cambridge, MA

Bailey N, Minton J (2018) The suburbanisation of poverty in British cities, 2004–16: extent, processes and nature. Urban Geogr 39(6):892–915

Cervero R, Day J (2008) Suburbanization and transit-oriented development in China. Transp Policy 15:315–323

Cervero R, Iuchi K, Suzuki H (2013) Transforming cities with transit: transit and land-use integration for sustainable urban development, urban development series. World Bank

Cervero R, Guerra E, Al S (2017) Beyond mobility: planning cities for people and places. Island Press, Washington, DC

Chang Y, Chen J (2013) Public housing in Shanghai: a tool with multiple purposes. Housing Finance International, Winter, pp 27–34

Chen F, Wu J, Chen X, Wang J (2017) Vehicle kilometers traveled reduction impacts of transit-oriented development: evidence from Shanghai City. Transp Res D 55:227–245

Day J (2013) Effects of involuntary residential relocation on household satisfaction in Shanghai, China. Urban Policy Res 31(1):93–117

Day J, Cervero R (2010) Effects of residential relocation on household and commuting expenditures in Shanghai, China. Int J Urban Reg Res 34(4):762–788

Feng J, Chen G (2018) Socio-spatial transformation of cities. In: Wu WP, Frazier MW (eds) The SAGE handbook of contemporary China. SAGE, Singapore

Fuchs C (2003) Structuration theory and self-organization. Syst Pract Action Res 16(2):133–167

Giddens A (1981) A contemporary critique of historical materialism. Vol. 1. Power, property and the state. Macmillan, London

Giddens A (1984) The constitution of society. Polity Press, Cambridge

Google Maps (2019) Shanghai maps. https://www.google.com/maps/place/Shanghai+People's+Square/@31.2292955,121.4722565,15.77z/data=!4m5!3m4!1s0x35b27069bae30e4f:0x69bed10757a4f1b8!8m2!3d31.22967!4d121.476161. Accessed 23 Feb 2019

Guan J, Xu C (2018) Are relocatees different from others? Relocatee's travel mode choice and travel equity analysis in large-scale residential areas on the periphery of megacity Shanghai, China. Transp Res A 111:162–173

He JH (2013) Implementation of the Shanghai master plan (2001–2020), Association of European Schools of planning 26th annual congress, 11–15 July. METU, Ankara

He S, Wu F (2005) Property-led redevelopment in post-reform China: a case study of Xintiandi redevelopment project in Shanghai. J Urban Aff 27(1):1–23

He S, Wu F (2007) Socio-spatial impacts of property-led redevelopment on China's urban neighbourhoods. Cities 24(3):194–208

Healey P (1997) Collaborative planning: shaping places in fragmented societies. UBC Press, Vancouver

Healey P (1998) Collaborative planning in a stakeholder society. Town Plan Rev 69(1):1–21

Hochstenbach C, Musterd S (2018) Gentrification and the suburbanization of poverty: changing urban geographies through boom and bust periods. Urban Geogr 39(1):26–53

Huang D, Huang Y, Zhao X, Liu Z (2017) How do differences in land ownership types in China affect land development? A case from Beijing. Sustainability 9(123):1–18

Jiading District Statistics Bureau (2014) Analysis on the survey results about living conditions of rural migrant workers in Jiading District, Jiading District, Shanghai Municipal Government. http://www.stats-sh.gov.cn/html/fxbg/201402/267398.html. Accessed 7 June 2018 (in Chinese)

Li H, Wei YD, Wu Y, Tian G (2019) Analyzing housing prices in Shanghai with open data: amenity, accessibility and urban structure. Cities 91:165–179

Liao B, Wong DW (2015) Changing urban residential patterns of Chinese migrants: Shanghai, 2000–2010. Urban Geogr 36(1):109–126

Liu CY, Ye L, Feng B (2018a) Migrant entrepreneurship in China: entrepreneurial transition and firm performance. Small Bus Econ 52:1–16
Liu Y, Fan P, Yue W, Song Y (2018b) Impacts of land finance on urban sprawl in China: the case of Chongqing. Land Use Policy 72:420–432
O'Donnel J (2018) The suburbanization of homelessness in Sydney, Australia. Appl Spat Anal Policy 11:333–354
Organization for Economic Co-operation and Development (2017) OECD economic surveys: China 2017, Paris, France
Pan HX (2006) Shanghai from dense mono-center to organic poly-center urban expansion, Urban Environmental Management Project of IGES, Department of Urban Planning, Tongji University, China
Pan H (2009) Rail transit and its impact on Shanghai metropolitan spatial structure, Second International Conference on Transportation Engineering 2009
Pan H, Shen Q, Liu C (2011) Transit-oriented development at the urban periphery: insights from a case study in Shanghai, China. Transp Res Rec 2245:95–102
Pan HX, Li J, Shen Q, Shi C (2017) What determines rail transit passenger volume? Implications for transit oriented development planning. Transp Res D 57:52–63
Pereira RHM, Schwanen T, Banister D (2017) Distributive justice and equity in transportation. Transp Rev 37(2):170–191
Peterson GE (2006) Land leasing and land sale as an infrastructure-financing option, World Bank Policy Research Working Paper 4043, World Bank
Randolph B, Tice A (2017) Relocating disadvantage in five Australian cities: socio-spatial polarisation under neo-liberalism. Urban Policy Res 35(2):103–121
Rawls J (1971) A theory of justice. Oxford University Press, Oxford
Sassen S (2001) The global city: New York, London, Tokyo. Princeton University Press, Princeton
Schuetz J, Giuliano G, Shin EJ (2018) Does zoning help or hinder transit-oriented (re)development? Urban Stud 55(8):1672–1689
Schwanen T, Lucas K, Akyelken N, Solsona DC, Carrasco JA, Neutens T (2015) Rethinking the links between social exclusion and transport disadvantage through the lens of social capital. Transp Res A 74:123–135
Sen A (1981) Poverty and famines: an essay on entitlement and deprivation. Clarendon Press, Oxford
Sen A (1984) Resources, values and development. Oxford, Basil Blackwell
Shanghai Metro (2019) Shanghai metro ticket. https://www.travelchinaguide.com/cityguides/shanghai/transportation/subway-ticket.htm. Accessed 23 Feb 2019
Shanghai Statistics Bureau (2012) Shanghai statistical yearbook 2012, Shanghai Municipal Government (in Chinese)
Shanghai Statistics Bureau (2017a) Shanghai statistical year book 2017, Shanghai Municipal Government (in Chinese)
Shanghai Statistics Bureau (2017b) Shanghai social and economic development report 2017, Shanghai Municipal Government (in Chinese)
Shen J (2011) Suburban development in Shanghai: a case of Songjiang, Thesis for the Degree of Doctor of Philosophy, School of City and Regional Planning, Cardiff University
Shen J, Wu FL (2013) Moving to the suburbs: demand-side driving forces of suburban growth in China. Environ Plan A 45:1823–1844
Shen Q, Chen P, Pan H (2016) Factors affecting car ownership and mode choice in rail transit-supported suburbs of a large Chinese city. Transp Res A Policy Pract 94:31–44
Shepard W (2016) How people in China afford their outrageously expensive homes, March 30, Forbes. https://www.forbes.com/sites/wadeshepard/2016/03/30/how-people-in-china-afford-their-outrageously-expensive-homes/#3795c7d3a3ce. Accessed 28 Dec 2018
Songjiang District Statistics Bureau (2014) Survey on living conditions of rural migrant workers in Songjiang District, Songjiang District, Shanghai Municipal Government (in Chinese)

Stanely J, Stanley J, Balbontin C, Hensher D (2019) Social exclusion: the roles of mobility and bridging social capital in regional Australia. Transp Res A Policy Pract 125:223–233

Tian L, Ge B, Li Y (2017a) Impacts of state-led and bottom-up urbanization on land use change in the peri-urban areas of Shanghai: planned growth or uncontrolled sprawl? Cities 60:476–486

Tian L, Guo X, Yin W (2017b) From urban sprawl to land consolidation in suburban Shanghai under the backdrop of increasing versus decreasing balance policy: a perspective of property rights transfer. Urban Stud 54(4):878–896

Tian L, Li Y, Shao L, Zhang Y (2017c) Measuring spatial-temporal characteristics of city expansion and its driving forces in Shanghai from 1990 to 2015. Chin Geogr Sci 27(6):875–890

Tongji University (2011) Residents travel demand analysis and transportation planning strategy in large-scale residential areas on the periphery of Shanghai Central City. School of Transportation Engineering, Tongji University, Shanghai

Wan G, Wang C (2015) Income polarization in the People's Republic of China: trends and changes, Asian Development Bank Institute, Working paper 538

Wang D, Gan D, Zhu C, Lu X (2012) A research on the relationships between suburban spatial planning and railway traffic planning in Shanghai. J City Plan 1:17–22. (in Chinese)

Wang Y, Feng S, Deng Z, Cheng S (2016) Transit premium and rent segmentation: a spatial quantile hedonic analysis of Shanghai metro. Transp Policy 51:61–69

Wu WP (2008) Migrant settlement and spatial distribution in metropolitan Shanghai. Urban Geogr 60:101–120

Wu Q, Li Y, Yan S (2015) The incentives of China's urban land finance. Land Use Policy 42:432–442

Yang S, Wang MYL, Wang C (2015) Socio-spatial restructuring in Shanghai: sorting out where you live by affordability and social status. Cities 47:23–34

You H (2016) Quantifying megacity growth in response to economic transition: a case of Shanghai, China. Habitat Int 53:115–122

Yuan J, Lin H (2013) An empirical study of commuting characteristics in rural areas. Procedia Soc Behav Sci 96:114–122

Zhang P, Zhou J, Zhang T (2017) Quantifying and visualizing jobs-housing balance with big data: a case study of Shanghai. Cities 66(2017):10–22

Zhang L, Yue W, Liu Y, Fan P, Wei YD (2018) Suburban industrial land development in transitional China: spatial restructuring and determinants. Cities 78:96–107

Zhou YX, Logan JR (2005) Suburbanisation of urban China: a conceptual framework, University at Albany

Chapter 7
Application of the Self-Organisation Approach in a Transport-Related Social Exclusion Model

7.1 Introduction

Among large cities in China (e.g., Shanghai, Guangzhou, Wuxi and Hong Kong), the conventional urban transport planning model has been widely adopted for urban transport planning (Wang et al. 2012; Pei and Fan 2010; Meyer and Miller 2001). The major goal of the model is to develop a plan for an efficient, balanced transport system for the rapid urban expansion of cities. The model involves four-step sub-models to project future traffic and is the basis for determining the need for new road capacity, transit service changes and land-use development. The model is based on transport systems, the characteristics of the population and the built environment to stimulate all travel behaviours of travellers.

However, cities are complex human social systems. Travellers' behaviours are the product of spontaneous and non-linear interactions among transport systems, individual decisions, the built environment and social rules (Fuchs 2003; Giddens 1984). The behaviours of travellers are difficult to predict using the conventional urban transport planning model, which is based on structural factors alone (Morrison 2005; Fuchs 2003). The transport literature finds that the conventional urban transport planning model has produced substantial differences between the forecasts of passenger travel behaviours and actual outcomes. The failure of the model to forecast real travel behaviours has resulted in wasting a huge amount of financial resources. Studies indicate that many city governments have built transport infrastructure that does not match their city's real travel demand (Andersson et al. 2017; Pan et al. 2011; Flyvbjerg et al. 2005).

To fill this research gap, this book introduces an innovative research approach that utilises the concept of the self-organisation process to investigate transport planning and social exclusion. The self-organisation approach focuses on low-income workers in deprived urban neighbourhoods whose travel behaviours are the product of an interaction between rules and individual decisions and actions.

J. C.-y. Lau, *Self-Organisation Shapes Travel Behaviours and Social Exclusion in Deprived Urban Neighbourhoods of China*,
https://doi.org/10.1007/978-981-15-2252-9_7

The travellers transform the rules and resources through their reproduction of routine practices to achieve their intended goals, such as reaching employment and recreational activities. The process produces emergence, including new travel behaviours, commuting problems, social exclusion and capability deprivation (Fuchs 2003; Giddens 1984). The self-organisation process provides new travel behaviours trends and travel demands, which seems to be a difficult task for conventional transport planning models to forecast and provide solutions for in terms of accessibility problems.

Structuration is the mechanism of the self-organisation process in human social systems (Giddens 1984). Structuration shapes the relationship between social rules and individual actions and also explains the process through which the rules are produced and reproduced by the actions of routine commuting practices to produce new travel behaviours. In addition, the concept of self-organisation helps to redefine the relationship among government policies, individual actions and travel behaviours. Travel behaviours are constituted by individual actions and structures (Giddens 1984). The transport literature fails to observe that individual actions are also an important factor in the production of new travel behaviours. The current transport literature is inadequate for understanding and using the relationships between social structures and individual actions to accurately forecast new travel behaviours.

This chapter also introduces a new transport-related social exclusion model, which is a self-organisation process that is shaped by the cycle of structure and agency. The model can help transport planners determine how the interaction between individual actions and structures produces new travel behaviours, including accessibility inequity, poverty and capability deprivation. The findings of the transport-related social exclusion model can also be incorporated into the conventional four-step transport-planning model to improve the accuracy of travel behaviour forecasts and minimise social exclusion.

Social exclusion obstructs low-income workers' rights and freedom to achieve well-being, and it takes place in the form of a multi-dimensional phenomenon. That is, low-income workers cannot achieve well-being because they experience resource deprivation, social inequity and a lack of freedom, choice and opportunity to participate in the mainstream activities of society (United Nations 2016; Sen 2000). It is unjust not only that government policies and social norms create capability deprivation but also that the workers' individual actions and decisions contribute to the creation of social exclusion. For example, the results of the studies in Chap. 5 indicate that economic and housing reforms have caused a concentration of low-income residents living in the deprived historic inner city of Xiguan. Some local residents have chosen to live in Xiguan because they want to live near job opportunities and near the social networks and recreational facilities that were developed in the inner city. Some have chosen to subdivide their houses into small rooms to accommodate more people, though this means that they have to live in an over-crowded environment.

Structure and agency are a duality; neither can exist without the other. Urban transport planners can improve the commuting of low-income people by changing either the structures or the individual practices (Giddens 1981, 1984). Therefore, this book suggests that transport planners use public consultation strategies to obtain information on individual actions to formulate appropriate new policies to facilitate individual actions (e.g., provide more freedom, choice and opportunity for poor residents to achieve well-being). In addition, decision-makers in global cities can change policies in accordance with social justice theories to minimise social exclusion and improve the accessibility of low-income workers. These two strategies can be incorporated into the proposed transport-related social exclusion model, which can be integrated into the conventional urban transport-planning model to improve the accuracy of the forecast and reduce social exclusion and inaccessibility problems.

The remainder of Chap. 7 is structured as follows. The next section introduces the concept of the self-organisation approach in urban transport planning and the measurement of personal accessibility by means of the concept of space-time prism. This section is followed by the introduction of the transport-related social exclusion model that is used to measure the accessibility of low-income workers. Then, there is a brief introduction to the conventional urban transport planning model and the incorporation of the results of the transport-related social exclusion model into the four-step model to improve the accuracy of travel behaviour forecasts and minimise social exclusion. The findings of the case studies in the previous chapters will be discussed to support the introduction of the above models. The chapter concludes with a discussion on the role of the self-organisation approach in urban transport planning.

7.2 The Concept of the Self-Organisation Approach in Transport Planning

In a complex human social system or a city, structures (e.g., traffic rules or transport systems) shape routine travel practices, all of which create varied patterns of behaviour. The individual actions and decisions of travellers lead to small changes and modifications in travel patterns. Interactions between structures and individual actions constitute social systems, including travel movements. In global cities such as Shanghai, daily work trips are being continuously produced and reproduced to create new travel behaviours. The process poses significant problems for planners in forecasting future travel behaviours in transport planning. The changing new travel behaviours create uncertainty in traffic forecasts and directly constrain transport planners in accurately predicting traffic demands and providing an appropriate supply of transit services for a region. Thus, transport planners need to develop a new approach to fill this research gap and improve the accuracy of travel demand forecasts. Based on the concept of the self-organisation approach, a new approach is introduced in this book to fill the research gap.

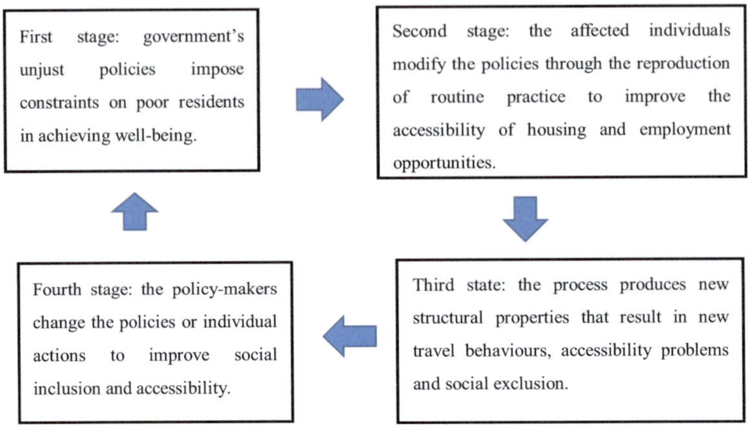

Fig. 7.1 The self-organisation process influences travel behaviour

The new approach is based on the self-organisation process and structuration theory, which states that social, land-use and economic structures are both the medium and product of travel practices (Giddens 1984). The literature on transport planning mainly focuses on the impacts of transport systems and the built environment on travel behaviour. However, travel behaviours are constituted by the interaction between structures and travellers' individual actions. Transport systems and the built environment are the medium for people's commutes, but at the same time, travel practices transform the built environment and transport systems to produce new travel behaviours that cannot be measured or estimated by conventional transport planning models. The self-organisation approach provides information on trends of new spontaneous and non-linear travel behaviours to help transport planners understand and forecast future traffic demands and provide solutions for transport problems. The self-organisation approach can be divided into four stages for analysing the trends of new travel behaviours, and the details are as follows:

The first stage concerns the influence of top-down emergences or government policies on the travel behaviours of the people in a region. In a human social system, the self-organisation process sometimes takes place in the form of a social exclusion process, in which unjust government rules interact with the agency of low-income workers to produce social exclusion problems. For example, the *hukou* system interacts with the actions of migrant workers and produces new structural properties that lead to the exclusion of the migrants from formal labour markets in urban China. When government rules like the above cause an imbalanced situation in a social system and reach a critical point, that influences the quality of life or even the survival of low-income workers. The effects of top-down emergence appear. The affected low-income workers will change their routine practices and find ways to achieve well-being (Rauws et al. 2016) (Fig. 7.1).

The second stage concerns bottom-up emergences, which occur when affected low-income workers carry out spontaneous self-organised activities to restore the balance of the system to achieve their intended goals, such as obtaining employment. The workers make use of their individual decisions and actions to transform the policies through the reproduction of their routine travel practices. For example, many low-income workers were trapped in the deprived inner city of Xiguan in Guangzhou during the economic and housing reforms. The residents subdivided their houses into rooms to accommodate more people, and they started up self-employed businesses in the inner city. In addition, the urban redevelopment and land finance policy displaced low-income residents from older urban neighbourhoods to suburban areas in Shanghai, and young displaced workers chose to travel long distances to find jobs. Bottom-up emergences produce new structural properties, and in this study, the new properties include the travel behaviours of low-income workers to reach affordable housing and informal employment. In addition, new travel behaviours lead to social exclusion problems, such as low-income workers restricting their daily activities to enclaves, resulting in commuting problems in reaching non-work activities (Rauws et al. 2016) (Fig. 7.1).

The third stage concerns the production of new structural properties by the self-organisation process, which changes the accessibility of the affected individuals in achieving their intended goals. In addition, emergence influences the accessibility of the fulfilment of other basic needs and directly adds uncertainty to transport forecasts. For example, in Chap. 4, the rural workers became concentrated in one area, and they depended on social networks to find informal jobs and affordable housing; these were the new structural properties that the *hukou* system and individual actions produced. At the same time, the workers' actions resulted in new travel behaviours, that is, they took short-distance trips, spatially restricted their social activities to the study urban village, spent a high number of time intervals on employment and only made social contact with their fellow villagers. These new behaviours caused accessibility problems in accessing social activities with local residents and other recreational activities (Fig. 7.1).

Emergences (or new structural properties) create new travel demands that cannot be estimated by conventional transport models. For example, in Chap. 6, the process of the suburbanisation of poverty in Shanghai produced emergences, including a large share of the displaced local workers being unable to find new jobs in the suburbs and choosing to take early retirement. These retirees adopted slow transport modes for reaching non-work activities. The new travel behaviours of the displaced workers contributed to the declining ridership of long-distance rail transit services.

The fourth stage concerns the modifications of policies or actions to produce new structural properties to improve the accessibility of low-income workers. City governments that are characterised by justice should seek ways to achieve accessibility equity. The term duality of structure indicates that the structural properties of social systems are both the medium and the outcome of the practices that constitute those systems. Structure and action are a duality; neither can exist without the other. Thus, policies and individual practices are inter-dependent variables; policy-makers can change either policy or individual actions to improve accessibility. Policy-makers can adopt social justice theories to deliver equitable policies for minimising social exclusion and improving accessibility. For example, Rawls (1971) proposed the difference principle, which states that transport policies are considered equitable if the policies give the greatest benefit to the least advantaged members, such as rural migrant workers (Fig. 7.1). Structure is capable of both constraining and enabling action (Giddens 1984). Healey (1997, 1998) proposed the collaborative approach, which helps policy-makers understand the attitudes and actions of all stakeholders by means of public consultation. The policy-makers determine appropriate policy responses to modify policies to minimise social exclusion and improve accessibility in the self-organisation process.

7.3 Measuring Travel Behaviour by Space-Time Prisms and Potential Path Area

The space-time prism approach is an important tool for measuring the personal accessibility and social exclusion of an individual in different stages of a self-organisation process. In each stage of the process, transport planners can use space-time prisms and potential path area (PPA) to determine any change in the travel behaviours (e.g., travel distances and travel times) of low-income workers. This section focuses on the space-time prism approach to measuring the accessibility of an individual in achieving well-being in the self-organisation process, exemplified by rural migrants' access to job opportunities and social and recreational activities. The space-time prism is based on the idea that the spatial extent to which a person can overcome space-time constraints to participate in desired activities can be measured by the size of his or her PPA. The shorter the distance a person commutes, the smaller the size of his or her PPA, the fewer the number of desired activities he or she can participate in, and thus, the lower the level of accessibility he or she experiences (Kim and Kwan 2003; Hägerstrand 1970). Therefore, in the chapters above, if rural migrant workers experienced short average travel times and adopted slow transport modes for work trips, they should be experiencing a shrinking PPA and thus be facing accessibility problems in accessing employment opportunities.

7.4 New Transport Planning Model: Transport-Related Social Exclusion Model

The purpose of this section is to use the concept of the self-organisation approach to focus on the investigation of the social exclusion and commuting problems of low-income workers. Between Chaps. 4 and 6, the self-organisation approach was used to determine the social exclusion and new commuting behaviour trends of low-income workers in urban China. This is a new approach in urban transport planning, and the author calls it the 'transport-related social exclusion model'. Furthermore, the research findings in the preceding chapters will be used to highlight the special characteristics of this new model in transport planning and social exclusion, as this model can provide information on new travel behaviour trends, as it seems to be difficult for the conventional urban transport planning model to forecast demands and provide solutions for transport problems. The findings of the new model can help transport decision-makers identify unjust policies and then deliver transformed policies to minimise social exclusion and inaccessibility problems. In addition, the findings on new travel behaviours can provide information on building appropriate transport infrastructure to match future travel demands. Not all of these tasks in the new model can be easily replaced by conventional urban transport planning models.

The research evidence in this book has indicated that social exclusion is a multi-dimensional phenomenon that takes place in urban China in the form of a self-organisation process. The mechanism of the process is the cycle of structure and agency, that is, structural factors (such as norms and government policies) interact with the individual actions of low-income workers to produce new travel behaviours or emergences that have resulted in the exclusion of individuals from participating in the mainstream activities of society. The transport-related social exclusion model includes five stages (Fig. 7.2).

7.4.1 The First Stage of the Model

The first stage concerns the structural factors that influence the commuting of low-income workers in their efforts to achieve well-being. Structural factors (e.g., policies) are the major causes of social exclusion (Sen 2000). Preston and Rajé (2007) suggest that social exclusion is not due to a lack of employment, education or opportunities for achieving well-being but rather to a lack of equitable government policies to ensure that poor residents have the freedom, availability of opportunity and choice to fulfil these basic needs. Structural factors influence individuals to achieve well-being and always take place through multi-dimensional phenomena. For example, the low-income rural migrant mothers discussed above do not have the freedom, choice and opportunity to participate in either non-work activities or employment mainly because they face an inequitable distribution of household duties due to social norms and the gendered division of household labour. The rural migrant women had to take on more household duties than the men, and thus, the former faced time conflicts between their jobs and household duties. Some had to find part-time jobs and earn low income. Using the findings of the surveys in the previous chapters, the author identified several structural factors that excluded low-income workers in urban China from achieving well-being; the details are as follows (Fig. 7.2):

- Chapter 4 concerned the impacts of government policies, such as the *hukou* system and land-use system, on the commuting and social exclusion of rural migrant workers and local villagers in the study urban village of Guangzhou. The

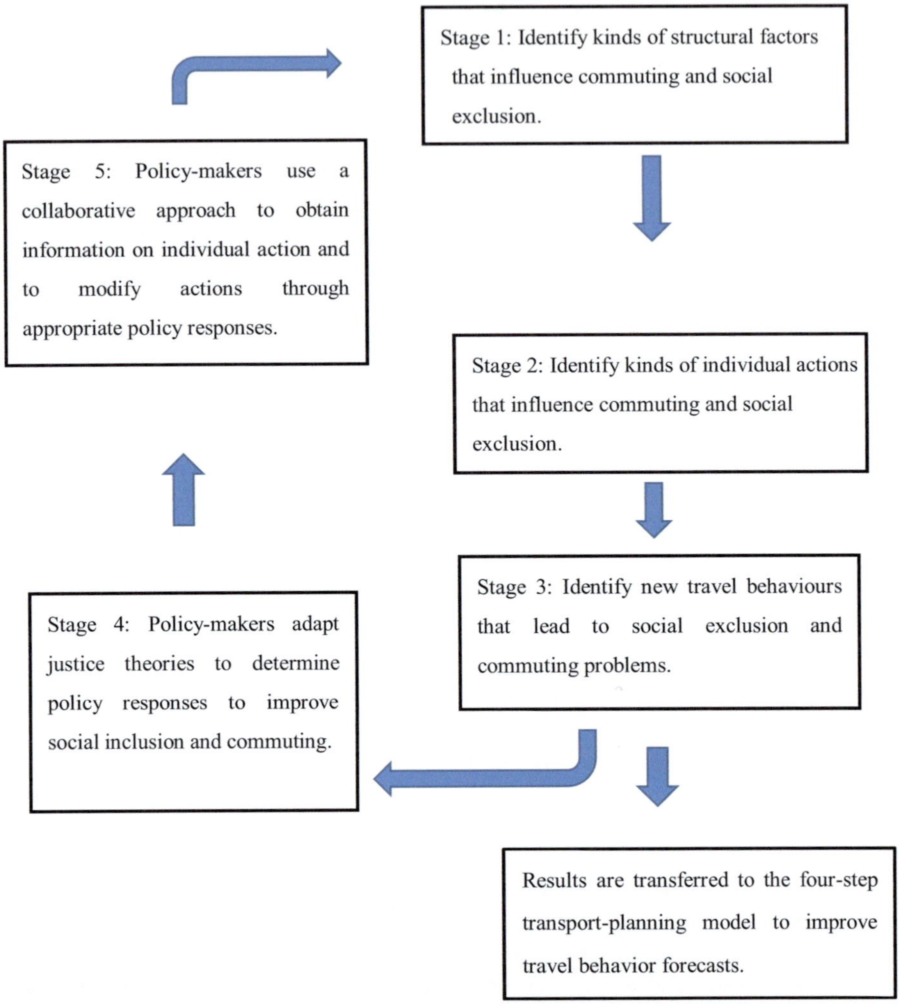

Fig. 7.2 Transport related social exclusion model

hukou system excluded rural migrants from participating in labour markets in urban areas. The concentration of rural migrants in the study urban village encouraged them to depend on social networks to find informal jobs and affordable housing in the urban village. The rural land-use system allowed the local villagers to build affordable housing and develop TVEs in the urban village. Both government policies, such as the *hukou* system and the dualistic land-use system, contributed to the exclusion of migrants and villagers from fulfilling basic needs. The rural migrants' social and employment activities were restricted to the urban village. They had to work long hours, earned low income and generated a lower level of trip frequency than the general working population of society. The villagers' farmland was acquired by the government, and the villagers built and leased houses and factory buildings to earn income, though that income was still low.

- Chapter 5 concerned the impact of economic and housing reforms on the exclusion of the low-income residents and migrants in the historic inner city of Xiguan. The local residents became redundant workers and were trapped, living in overcrowded and dilapidated houses. The *danwei* system influenced the travel behaviour of the residents, some of whom still rode bicycles and walked to work. Because of the *hukou* system, the migrant workers in the inner city earned lower wages than local workers and did so in substandard conditions.
- Chapter 6 concerned the impact of the suburbanisation policies in Shanghai on the exclusion of displaced local residents and rural migrants. Many of the displaced local residents were involuntarily moved from the city centre to the suburbs because of the urban redevelopment and land finance policies. Most of the residents were low-income, poorly educated and more advanced in age and thus experienced difficulties in seeking new jobs in the suburbs. A large share of them became unemployed and chose to retire early. In addition, the acquisition of farmland and the urban sprawl in Shanghai caused the dispersion of migrants' housing and employment locations. Migrants' social and employment activities were restricted to migrant enclaves, some of which were located far away from rail transit networks. As a result, most displaced workers and migrants took short-distance trips and used slow transport modes to reach activities.

7.4.2 The Second Stage of the Model

The second stage involves identifying the impacts of the individual actions and attitudes of low-income workers on their commuting and exclusion. Social exclusion always occurs in deprived urban neighbourhoods, where the individual actions of low-income workers interact with various social norms and unjust government policies to produce new structural properties. Human actors are not passive objects. They make use of individual decisions and actions to transform social norms and policies to achieve their intended goals, such as obtaining employment and housing. Using the results of the surveys in the previous chapters, the author has identified several kinds of individual actions and decisions that are related to the social exclusion of the low-income workers; the details are as follows (Fig. 7.2):

- Chapter 4 was mainly about how the individual actions of rural migrants interacted with the *hukou* system in the search for jobs and housing in the study urban village. Migrants moved swiftly from the urban districts in Guangzhou to live in the study urban village. They found manufacturing jobs that were provided by the township and village enterprises (TVEs). They depended on social networks to find jobs and housing, worked long hours to earn an adequate income and generated a lower level of trip frequency than the general working population.

The village collectives made use of the dualistic land-use system to develop TVEs in order to tap into the cheap labour sources provided by rural migrants. The actions of rural migrants and villagers helped them achieve their intended goals, such as local villagers leasing their housing and factories to earn income and migrant workers finding jobs and housing. However, both migrants and local villagers restricted their social and employment activities to the study urban village.

- Chapter 5 concerned the impacts of the individual actions of low-income local residents and rural migrants on the exclusion of residents in Xiguan, Guangzhou. The economic and housing reforms increased the poverty of the local residents and trapped them in the deprived area of Xiguan. They subdivided their houses into small rooms to accommodate more people. They started up self-employed retail businesses in Xiguan, including low-end commerce, wholesale, and retail-based consumption service jobs. They developed strong social networks to help each other find jobs. Many of them walked, rode bicycles and took buses to work, which was located near their homes. The rural migrant workers chose to gather on the western side of Xiguan and developed social networks to help them find wholesale clothing and food catering jobs. They also used slow transport modes to reach activities. Local residents, who had to work long hours, took short-distance trips and generated a lower level of trip frequency than the general working population.
- Chapter 6 was about the individual decisions and actions of the displaced local workers and migrants that influence their exclusion and commuting in the suburbanisation process in Shanghai. Many migrant workers choose to rent private housing to live near manufacturing and retailing job opportunities in the inner suburbs. They borrow money from relatives to start self-employed businesses and have developed enclaves where the concentration of migrants is based on their provinces of origin. Gradually, their actions have transformed the enclaves into business and retail centres in the new towns. Their social and recreational activities are restricted to the enclaves, and they thus take short-distance trips and adopt slow transport modes to each activities.

  Many of the displaced local residents in the suburbs are low-income, poorly educated and elderly workers. Most of the young displaced residents choose to travel long distances from new towns to find job opportunities in the city centre. A large share of middle-aged and local workers with low levels of education have failed to find new jobs in the suburbs, and they have chosen early retirement. They travel short distances for non-work trips and use slow transport modes.

7.4.3 The Third Stage of the Model

The third stage aims to identify new travel behaviours that are produced by the self-organisation process and are the cause of inaccessibility and exclusion. Social exclusion is a multi-dimensional phenomenon. In the previous chapters, this book introduced the idea that the impacts of new travel behaviours on low-income workers can be broadly classified into three categories, including new travel behaviours that cause the poverty of poor workers, those that lead to capability deprivation in achieving well-being and those that lead to accessibility inequity (United Nations 2016; Sen 1984, 2000; Rawls 1971). Using the findings of the case studies in Chaps. 4–6, the author has identified several new travel behaviours that were produced by the self-organisation processes and resulted in transport-related social exclusion. The details are as follows (Fig. 7.2):

- Chapter 4 was about how the rural migrants' actions interacted with the *hukou* system to produce new travel behaviours, including their employment opportunities and social networks being spatially restricted to the study urban village. The migrants had low incomes and could not afford to adopt fast transport modes to reach work. They generated short-distance work and social trips that produced accessibility inequity. The influences of social networks, the co-location process between jobs and housing and the gendered division of household labour deprived many of the migrants from having adequate time intervals to travel to distant recreational and social trips in other districts of Guangzhou. In addition, rural migrants faced capability deprivation, that is, they were obstructed from having the freedom, choice and opportunity to commute to public housing, education and medical services.
- Chapter 5 concerned the new behaviours produced by low educational attainment and the economic and housing reforms interacting with the individual actions of local residents in Xiguan, Guangzhou. The new structural properties included the local residents being unable to afford to move out and having to find jobs in the inner city. They engaged in informal retail jobs in the inner city and lived in poverty. They faced accessibility inequity because their commuting was constrained by inadequate public transit services and outdated road systems. They had to work long hours to earn adequate income and thus had less time and fragmented spare time for commuting to distant social activities; therefore, they generated a lower level of trip frequency than the general working population of the society and generated short-distance trips. The local residents were deprived of the freedom, choice and opportunity to commute to mainstream activities in the city.

  In the face of keen competition over employment opportunities, migrant residents in Xiguan developed strong social networks to avoid the impacts of stigma from local residents and the disruption of their social relationships. The migrants developed an enclave that included the walking streets and wholesale clothing markets on the western side of Xiguan. As a result, they also travelled short distances for employment and social activities. They earned low income and were deprived of the freedom and opportunity to reach employment, social and recreational activities in other districts of Guangzhou.

- Chapter 6 concerned the new behaviour produced by the self-organisation process in the suburbanisation of Shanghai. The new behaviour includes many of the residents facing accessibility inequity, such as jobs-housing spatial mismatch accessibility problems and inadequate public transit services in the suburbs. Many of them have chosen to retire early and thus live in poverty. They take short-distance trips and have adopted slow transport modes to reach activities. They were deprived of the freedom and choice to commute to jobs and social activities in the city centre.

  The migrant workers in inner suburbs mostly generate short distance social and work trips. They are deprived of the freedom to commute make social contact with local residents and to commute to distant recreational and employment opportunities. The new travel behaviours of migrants and local residents have resulted in the declining ridership on rail transit services.

7.4.4 The Fourth Stage of the Model

The fourth stage is to identify and modify unjust policies in the self-organisation process to minimise social exclusion and improve the commuting of low-income workers. City governments that seek to implement the characteristics of social justice should try to formulate new transport policies to improve accessibility equality. Using the findings of the studies in the preceding chapters, the author proposes that social justice theories be used to improve the social inclusion of low-income workers. The details are as follows (Fig. 7.2):

- In Chap. 4, the author suggested that the Chinese government transform the *hukou* system in accordance with the difference principle (Rawls 1971), which states that inequity in the distribution of resources is permissible only if it benefits the people in the least well-off positions of society, such as rural workers in urban China. Therefore, the *hukou* system should be modified so that rural migrants can be included in the formal labour markets and social welfare services of host cities. The inclusion of migrants could increase their freedom, opportunity and choice in obtaining employment, which would allow them to commute longer distances and use faster transport modes for work trips than their existing travel behaviours include. The change would also increase their accessibility to non-work activities.
- In Chap. 5, the author suggested the capability approach (Sen 1980, 1984, 1992) to improve the housing and job market environment in the historic inner city of Xiguan. The capability approach stresses that all residents in Xiguan are entitled to the freedom and opportunity to achieve well-being. Thus, if the housing and economic policies are ere amended, the local residents can be provided with new public housing with adequate basic facilities. In addition, the government should provide adequate formal job opportunities for local residents so that they have the

freedom to choose employment either inside or outside of Xiguan. In addition, the changes would increase their trip distances, and they would use faster transport modes for work and non-work trips, resulting in improved accessibility.

- In Chap. 6, the author suggested that the Shanghai government transform its suburbanisation policies so that the policies are of the greatest benefit to the disadvantaged, such as displaced low-income local workers having the freedom to choose their housing location (Rawls 1971). Therefore, if the land finance and urban redevelopment policies are amended in the process of suburbanisation, the displaced local workers can live close to job opportunities and rail transit lines in suburban areas. In addition, policies should be amended so that there are adequate employment opportunities to match the skills of the relocated workers. The changes would increase the workers' trip distances, and they would use faster transport modes for work and non-work trips, resulting in improved accessibility. The change in policies would also increase the ridership of the rail transit services in the suburbs.

7.4.5 The Fifth Stage of the Model

The fifth stage of the model aims to identify and change individual actions in the self-organisation process by means of the collaborative approach. Policy-makers can obtain information on individual actions and try to formulate appropriate policies to facilitate those individual actions and improve social inclusion. The collaborative approach was originally used to guide the organisation of political communities to obtain the views and attitudes of all stakeholders on government policy (Healey 1997). In this book, the collaborative approach is adopted as a strategy for public consultation to understand the individual actions of shareholders, minimise social exclusion, improve accessibility and produce new travel behaviours. Lucas (2012) suggests that policy-makers can minimise social exclusion if they recognise the abilities, skills, resources, capacities and experiences of affected individuals in the design of policy solutions. This concept is in line with the duality of structure, which states that policy is both the medium and the product of action. Policy and action are a duality; neither can exist without the other. Policy-makers can change the social exclusion process by changing structures or actions (Giddens 1984) (Fig. 7.2).

Collaborative planning is a procedure used to obtain all stakeholders' views on the process of public consultation. It should be interactive and will involve many of those with a stake in the policy changes to improve social inclusion and commuting (Healey 1998). The government cadres in China should respect the rights of all stakeholders to voice their views on government policies. The collaborative approach, which includes two types of institutional infrastructure, namely, hard (e.g., street offices and district governments) and soft (e.g., radio and television programmes) infrastructure, requires government cadres to break out of traditional hierarchical and 'bureaucratic' processes to involve new groupings and networks.

7.5 Incorporating the Transport-Related Social Exclusion Model Into the Conventional Urban Transport Planning Model

This book argues that the neighbourhood effect, like the self-organisation process that is taking place in the neighbourhood, influences residents' travel behaviours. In addition, the self-organisation processes in the deprived neighbourhoods are significantly different from those in the high-income neighbourhoods of a city. Therefore, if the city governments use the aggregated transport-planning model to estimate travel behaviours of their cities, the travel behaviours of the working residents in the deprived neighbours will be obscured, and the city governments will be unable to provide suitable transport systems for these deprived neighbourhoods, resulting in inaccessibility and social exclusion. In fact, the results of the studies in Chap. 6 indicate that the declining ridership of the rail transit services in the suburbs is mainly due to the travel behaviours of the working residents in suburban areas in Shanghai not matching the forecasts of the transit-oriented development and conventional urban transport planning model (Pan et al. 2011; Wang et al. 2012).

The results of the studies in this book found significant relationships between several dimensions of the neighbourhood context and the commuting behaviours of low-income residents. In the deprived neighbourhoods of Guangzhou and Shanghai, there were links between the neighbourhoods and social structures (such as government policies, norms, and social networks) and individual decisions and actions. In the deprived neighbourhoods, the interaction between structure and agency appeared to significantly obstruct the low-income residents' freedom and choice in achieving well-being and negatively influenced their travel distances, trip frequency and mode choice to reach the locations necessary to fulfil their basic needs.

This book suggests that the impact of the neighbourhood effect (e.g., the interaction between government policies and the individual actions of residents) on the individual commuting behaviours of low-income workers may be obscured in conventional urban transport planning models that measure the aggregated travel patterns of people across all districts in a city. Therefore, by focusing on the deprived urban neighbourhoods, this study can propose a model to integrate the transport-related social exclusion model (which measures the individual travel behaviours of poor residents in deprived neighbourhoods) into the conventional urban transport planning model to improve transport forecasting and minimise the social exclusion of low-income workers.

7.6 The Travel Behaviours of Low-Income Workers in Deprived Neighbourhoods

According to the results of the studies in Chaps. 4 through 6, the low-income workers' new travel behaviours that were produced by the social exclusion processes in the deprived neighbourhoods of Guangzhou and Shanghai are as follows:

- In the inner city of Xiguan, because of the revitalisation of the economy, many low-skilled local residents could depend on social networks to find jobs. Most of the employed adults found jobs along the tourist streets and thus restricted their employment to Xiguan.
- Because of the high rate of residential mobility and the individual actions taken to find jobs, rural migrant workers in the study urban village and Xiguan tended to live in proximity to job opportunities.
- The long time intervals spent working interacts with the decisions of rural migrant workers to earn more income, so most of them choose to take short-distance trips.
- In Shanghai, displaced middle-aged local workers often choose early retirement and take short-distance trips to reach non-work activities.
- As a result of the interaction between the gendered division of household labour and individual actions, low-income working mothers choose to seek jobs in proximity to their homes to minimise travel times and reduce time conflicts between their household duties and employment;
- Because of the co-location process, local villagers in the study urban village built factories in proximity to the homes of rural migrant workers in order to tap into the cheap labour source, and this process led to reduced travel times.

All of the low-income workers in the deprived neighbourhoods generated low work trip rates because most of the time intervals were spent at their workplaces. They used slow transport modes to reach activities in order to reduce transport costs.

7.7 The Conventional Four-Step Transport Planning Model

This section introduces the conventional four-step transport-planning model, which has been widely used in large cities, including Shanghai, Beijing and Guangzhou in China. The four steps of the model are briefly introduced below (Bliemer et al. 2017; Andersson et al. 2017; Meyer and Miller 2001; McNally 2000) (Fig. 7.3).

1. The trip generation step forecasts the total number of home-based and non-home-based trips produced from the points of origin and attracted to the destinations in the traffic analysis zone. The trip rates between origins and destinations are predicted in order to determine the total amount of trip generation. The total trip generation is influenced by the proximity among activity locations, the

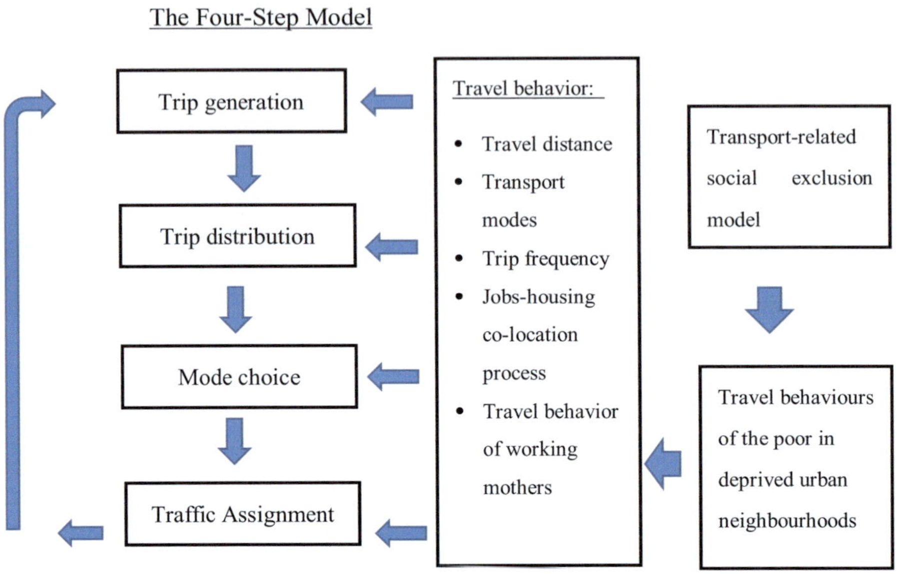

Fig. 7.3 Incorporation of the results of the transport-related social exclusion model into the four-step transport planning model

efficiency of the transport system and the demographic features of the population in the traffic analysis zone.

2. The trip distribution step forecasts the origin-destination patterns future trips affected by population, employment and other demographic changes that reflect changes in people's choice of destination. Travel distances determine the trip rates between origin-destination locations. The closer the distances between origins and destinations are, the higher the rate of trips the travellers will make. Therefore, the gravity model is used to forecast trip distribution patterns. The gravity model assumes that travel distance is the major friction factor for travelling, and thus trip frequency declines with increasing distance between production-attraction pairs.
3. The mode choice step transfers the trip tables from trip distribution to produce mode-specific trip tables. This step predicts the proportion of a particular transport mode in a particular traffic path flow in the traffic analysis zone. This model depends on the availability, distance, and travel times of each respective transport mode as well as the preferences of travellers. For example, in the case of short-distance commuting, travellers will choose slow and low-cost transit services. The mode choice model helps transport decision-makers predict the demand for a particular transport mode on a particular traffic path flow so as to supply appropriate transit services for low-income workers.
4. The traffic assignment step loads the data generated in the origin-destination trip matrix and mode choice table onto the transportation network. The traffic assignment sub-model spreads the traffic path flows and mode choice data through the

network and produces travel times. It is assumed that the decisions of travellers are influenced by the infrastructure characteristics, including the road segment length, number of lanes and maximum speed in a traffic analysis zone (Bliemer et al. 2017). Often, Wardrop's principle of user equilibrium is applied, which means that each driver chooses the trip with the shortest travel time. If the predicted traffic flows exceed the capacity of the specific road network (e.g., rail transit services), congestion may occur, which may negatively affect travel time. The traffic assignment step is used as a reference for governments' transport infrastructure investments to meet increasing traffic volume.

The transport literature has found that there are uncertainties and substantial differences between travel volumes forecast by the four-step model and the actual outcomes. A very large number of traffic forecast errors can be explained by the input variables being different in reality from what is assumed in the forecasts; for example, the trip frequency, mode choice and travel distance of rural migrant workers are substantially different from the travel behaviour of the general population in urban China (Andersson et al. 2017). This book argues that travel behaviour is the product of the interaction between structure and agency; therefore, the input variables for the four-step model should also consider the results of the transport-related social exclusion model that produces new travel behaviours in terms of trip frequency, travel distance and mode choice, in order to improve the accuracy of forecasts (Fig. 7.3).

7.8 Incorporation of the Results of Transport-Related Social Exclusion Model in the Trip-Generation Step

The trip generation step focuses on the forecasts of the total trip production, including those trips that are produced in the original locations and those trips that are attracted to the destination locations in the study zone. The total trip production is also related to the frequency of trips between origins and destinations in the study zone. The forecasts ignore many causal factors that contribute to the generation of trips, such as government policies and social norms that interact with the individual decisions and actions of the people in the study zone.

The results of the studies in Chap. 4 indicate that the low-income workers in the study urban village in Guangzhou chose to work long hours to earn an adequate income, generated a low level of trip frequency, chose to make social contact with their fellow villagers instead of local residents in the host cities, and found jobs by means of social networks; thus, their daily work trips were restricted to the migrant enclave. In addition, Chap. 5 examined how the low-income local residents in the historic inner city of Guangzhou chose to subdivide their houses into small rooms to accommodate more people and depended on their social networks to find jobs. The local residents chose to live and work in the inner cities instead of relocating to the suburbs and finding jobs in suburban districts. The results indicate that it is neither the built environment nor the transport system that influences the generation of trips; rather, the decisions of the low-income workers interacted with government policies with the result that most of the workers' trips take place in the deprived neighbourhoods and that the workers mostly generate short distance trips, adopt slow transport modes and generate lower levels of trip rates than the general working population in Guangzhou.

Hence, to improve the accuracy of travel demand forecasts, transport planners should understand the trends in the new travel behaviours of these low-income workers in deprived urban neighbourhoods. The planners should conduct qualitative surveys in deprived urban neighbourhoods that are based on transport-related social exclusion models to confirm travel distances, trip rates and mode choices between origins and destinations. In addition, the transport planners, who are responsible for the forecasting, should transfer the results of the surveys in the deprived neighbourhoods to the trip generation model. The planners must amend the input variables (e.g., adjust the variables of origin and destination, travel distance, trip frequency and mode choice in the deprived neighbourhoods) to improve the accuracy of trip generation forecasts (Fig. 7.3).

7.8.1 Incorporation of the Results of the Transport-Related Social Exclusion Model Into the Trip Distribution Step

The trip distribution step is based on the idea that trip frequency distributions are mainly influenced by travel distances between the origins and destinations. Thus, trip frequency declines when the spatial distance between the origin-destination pair increases. In addition, the travel distance between origins and destinations is mainly influenced by the proximity between locations, the performance of transport systems and the income of the travellers. The results of the studies in this book indicate that the structural factors of government policies and social networks influence low-income workers' travel distance, mode choice and trip frequency to reach employment and social activities.

For example, in Chaps 4 and 6, because of the *hukou* system, revitalisation policy in Xiguan and social networks in the enclaves, the low-income workers chose to live near job opportunities and depended on social networks to find jobs. As a result, their travel distance to work was not influenced by the proximity between the locations of major employment centres and their homes. The results of the studies in Shanghai found that only a small share of low-income workers who live in the new towns have sought jobs in the major employment areas in the city centre. Many rural workers in the studies in Guangzhou and Shanghai exhibited a high rate of residential mobility and moved swiftly to live near informal job opportunities. Therefore, as far as the trip distribution step is concerned, in the deprived neighbourhoods, a large share of the low-income workers' trips are short-distance trips, and the workers take slow transport modes and generate a low trip frequency to reach employment and social activities. These behaviours are significantly different from the travel behaviours of the general population, whose trip distributions are mainly dependent on the transport system, the built environment, and the proximity among locations, such as the proximity between major employment centres and housing locations.

Transport decision-makers should consider these factors before they make the decision to invest large-scale financial resources to build transport infrastructure, such as changing the scale of rail transit services for the deprived neighbourhoods in suburban areas. For example, if the decision-makers had considered the self-organisation process in the deprived neighbourhoods as they planned the suburbanisation of Shanghai, the case of declining rail ridership in the suburbs could have been avoided.

The author suggests that the transport planners should first conduct household surveys in the deprived urban neighbourhoods that are based on the transport-related social exclusion model. After the planners have discovered the new travel behaviours or emergences that are produced by the self-organisation process, the transport planners should amend the input variables (e.g., travel distance, trip frequency and mode choice in the deprived neighbourhoods) in the trip distribution step (Fig. 7.3).

7.8.2 Incorporation of the Results of the Transport-Related Social Exclusion Model Into the Mode Choice Step

The mode choice step is used to transfer the trip tables from trip distribution to produce mode-specific trip tables. This step can reflect a range of transport system performance variables, such as the availability of each transport mode their respective travel times and distances and trip-maker characteristics. However, the mode choice step fails to consider the influence of individual actions and structural factors, such as social networks and government policies.

This section has argued that structural factors alone, e.g., transport systems and the built environment, cannot fully explain new travel behaviours. According to the self-organisation process, the interaction between structural factors and individual actions produces new travel behaviours, including the mode choices of low-income workers.

For example, Chap. 5 described how many local residents in the historic inner city of Guangzhou were accustomed to riding bicycles to work and non-work activities, a habit they had developed in the *danwei* era (Zacharias 2012). In addition, in Chap. 6, which investigated the suburbanisation of poverty in Shanghai, the Shanghai Master Plan (1999–2020) proposed building a citywide transport plan to support the suburbanisation plan, and each important town was connected by rail. The transport planners predicted that a large share of the relocated population in new towns would travel long distances to seek jobs in the city centre (Pan 2006). According to the real travel behaviours of the low-income residents in Shanghai, most of them actually take short-distance trips and adopt slow transport modes because of the influences of the *hukou* system, urban redevelopment policies and social networks. Furthermore, transport planners had followed the design guidelines of transit-oriented development (TOD) when they planned and built the rail transit stations. The medium densities of residential and retail development were developed around suburban rail station areas. However, the real travel behaviour was that many relocated workers in the suburbs did not take rail transit services to work (Guan and Xu 2018; Pan et al. 2011).

Obviously, the transport planners in Shanghai used the four-step model to forecast the trip generation, trip distribution, mode choice and traffic assignment of travellers at the beginning of the suburbanisation plan (Wang et al. 2012). Since there were many job opportunities in the city centre, the model predicted that the home-based trips produced from the residential areas in new towns would be connected to the major job areas in the city centre. Based on the four-step model, the government invested huge financial resources in building a rail network and developed TOD to link homes and job centres.

The four-step model failed to observe that the rural migrants accounts for more than 50% of the total population in the inner suburbs, and the rural migrants' travel behaviour is not mainly influenced by the built environment, jobs and housing locations or transport systems (Shanghai Statistics Bureau 2017). The travel behaviour of rural migrants is significantly influenced by the interactions between social networks, the *hukou* system and individual actions. The migrants choose to live and work in migrant enclaves. They travel short distances as part of their work obligations and take slow transport modes to reach activities. The four-step model failed to forecast the travel behaviour of the rural migrants, as it failed to forecast that the rural migrants' trips would be concentrated in the inner suburbs and involve slow transport modes. As a result, the rural migrants, who account for a major proportion of the population in the inner suburbs, cannot provide adequate travel demand for rail transit services.

If the transport planners in Shanghai had adopted the transport-related social exclusion model, they would have avoided over-estimating the rail traffic demand in the suburban areas (Fig. 7.3). The transport planners should first conduct surveys in migrant enclaves or deprived neighbourhoods that are based on transport-related social exclusion models. After they have discovered new travel behaviours or emergences, they should amend the input variables accordingly (e.g., adjust the variables of origin and destination table, travel distance, trip frequency and mode choice in the deprived neighbourhoods) and then transfer the input variables into the mode choice model to improve the accuracy of rail demand forecasts (Fig. 7.3).

7.8.3 *Incorporation of the Results of the Transport-Related Social Exclusion Model Into the Traffic Assignment Step*

The traffic assignment step is used to forecast the route choice and mode choice of travellers in accordance with the origin-destination trip matrix. The step also spreads the traffic path flows and mode choice data through the network and produces travel times. It is assumed that the decisions of travellers are influenced by infrastructure characteristics, including the road segment length, number of lanes and maximum speed (Bliemer et al. 2017).

In the traffic assignment step, transport planners assume that the characteristics of transport infrastructure and the availability of transport modes determine travel times and thus determine the distribution of traffic flows. However, the results of the studies in the preceding chapters indicate that the self-organisation process always has greater influence on travel time, travel distance and mode choice than the built environment and transport infrastructure do, as seen with the commuting patterns of rural migrants and local villagers in the study urban village in Chap. 4. Due to the influence of the *hukou* system, the migrants moved swiftly from other districts and gathered in the study urban village to find informal manufacturing jobs. Furthermore, the dual land-use system allowed the village collectives to develop TVEs on rural land to tap into the cheap migrant labour force. The interactions between structures and individual actions produced a jobs-housing co-location process to cut travel times on work trips. The daily trips of the migrants were restricted to the study urban village.

The author suggests that transport planners first conduct household surveys in the deprived urban neighbourhoods that are based on the transport-related social exclusion model. After they have discovered the new travel behaviours or emergences that are produced by the self-organisation process, transport planners should amend the input variables accordingly (e.g., adjust the variables of the route choice table, origin and destination table, travel distance, trip frequency, and mode choice) before they are input into the traffic assignment model for travel behaviour forecasting (Fig. 7.3).

7.9 Conclusion

This section concludes the findings of the studies in the preceding chapters of the book through the introduction of the new transport-related social exclusion model for urban transport planning. The concept of the self-organisation process influencing the commuting of poor residents is applied to the social exclusion and conventional urban transport planning models to improve the forecasting of travel behaviours in deprived urban neighbourhoods.

This chapter first introduced the self-organisation approach, which is based on the idea that the human social system can be self-organised to achieve a balanced environment. Human actors followed social rules and resources to organise social activities. The actors transformed the rules to achieve their intended goals, such as reaching employment and social activities. The self-organisation process produced new travel behaviours that led to accessibility problems and social exclusion. By understanding the self-organisation approach, transport planners can understand and resolve commuting problems and minimise social exclusion.

Using the results of the studies in the preceding chapters, this chapter noted that commuting behaviours are influenced not only by the built environment, transport system and main socio-economic characteristics of the population but also by the actions and decisions of individuals and other structural factors. The individual decisions and actions interact with structural factors (e.g., transport rules) to produce new travel behaviours, which are called emergences; transport planners have been unable to predict these emergences using the conventional transport planning model or structural factors alone.

In addition, the self-organisation concept can be used to investigate social exclusion in the deprived neighbourhoods in urban China. Unjust policies interact with individual decisions to produce social exclusion; for example, the *hukou* system interacts with rural migrants' actions in migrant enclaves. Hence, to resolve accessibility and social exclusion problems, government decision-makers must understand the cycle of structure and agency that produces the accessibility constraints that obstruct the low-income workers' freedom, choice and opportunity in fulfilling their basic needs.

Social exclusion is a multi-dimensional self-organisation process, and this concept can be applied in transport planning through the new transport-related social exclusion model. To understand complex travel behaviours and social exclusion, transport decision-makers must conduct surveys to obtain information on the interaction between structural factors and individual actions in the deprived urban neighbourhoods. The self-organisation process produces new structural properties that obstruct the low-income workers' freedom, choice and opportunity to achieve well-being. The process also produces poverty and accessibility problems. Improving this process includes five stages. The first stage is to identify the influence of structural factors on individuals. The second stage is to identify individual actions that interact with structural factors to produce new travel behaviours. The third stage is to identify the new travel behaviours that result in the exclusion of low-income workers from participating in the mainstream activities of society.

The fourth stage is to adopt social justice theories to modify unjust policies and deliver equitable policies for minimising social exclusion and improving accessibility. The final stage is to use the collaborative approach to understand the individual actions and attitudes of actors so as to modify unjust policies, facilitate the actions of low-income workers, and minimise social exclusion and inaccessibility. In introducing the different stages of the transport-related social exclusion model, the results of the case studies between Chap. 4 and 6 were used to support the concepts.

According to the concept of the neighbourhood effect, the commuting behaviours and social exclusion phenomenon in deprived urban neighbourhoods should be considered differently than those in high-income neighbourhoods in the urban transport planning of a city. This is because the self-organisation processes between these two types of neighbourhoods are different. In urban transport planning that uses the aggregated measures of travel behaviours, the transport behaviours of low-income workers in deprived neighbourhoods can be obscured. Thus, using this type of planning, the city governments cannot provide appropriate transport services for the deprived neighbourhoods. To fill this research gap, this chapter introduced an approach to incorporate the transport-related social exclusion model into the conventional four-step urban transport planning model.

The four-step model, which has been widely used in large cities, mainly estimates the impacts of transport infrastructure, the built-environment and the main socio-economic characteristics of the population on travellers' travel behaviours, including trip generation, trip distribution, mode choice and traffic assignment. The four-step model fails to forecast the travel behaviours in deprived urban neighbourhoods, where the travel behaviours are produced by the interactions between structures and the individual actions of low-income residents. To improve the accuracy of travel behaviour forecasts in the deprived urban neighbourhoods, this chapter proposed that the transport planners who are responsible for future planning should conduct individual surveys in in the deprived neighbourhoods that are based on the concept of the self-organisation process. After they obtain information about the new travel behaviours of the low-income residents, (such as taking short-distance trips, using slow transport modes and generating a low level of trip frequency to reach activities), the planners should incorporate the new travel data into the conventional four-step transport planning model to estimate the travel demand in deprived urban neighbourhoods. It is suggested that if the transport decision-makers in Shanghai had used this new transport planning method, the declining ridership of rail transit services in the suburbs would have been avoided.

References

Andersson M, Brundell-Freij K, Eliasson J (2017) Validation of aggregate reference forecasts for passenger transport. Transp Res A 96:101–118

Bliemer MCJ, Raadsen MPH, Brederode LJN, Bell MGH, Wismans LJJ, Smith MJ (2017) Genetics of traffic assignment models for strategic transport planning. Transp Rev 37(1):56–78

Flyvbjerg B, Holm MKS, Buhl SL (2005) How (in)accurate are demand forecasts in public works projects? J Am Plan Assoc 71(2):131–146

Fuchs C (2003) Structuration theory and self-organization. Syst Pract Action Res 16(2):133–167

Giddens A (1981) A contemporary critique of historical materialism. Vol. 1. Power, property and the state. Macmillan, London

Giddens A (1984) The constitution of society. Polity Press, Cambridge

Guan J, Xu C (2018) Are relocatees different from others? Relocatee's travel mode choice and travel equity analysis in large-scale residential areas on the periphery of megacity Shanghai, China. Transp Res A 111:162–173

Hägerstrand T (1970) What about people in regional science? Reg Sci Assoc Pap 24(1):7–21

Healey P (1997) Collaborative planning: shaping places in fragmented societies. UBC Press, Vancouver

Healey P (1998) Collaborative planning in a stakeholder society. Town Plan Rev 69(1):1–21
Kim HM, Kwan MP (2003) Space-time accessibility measures: a geo-computational algorithm with a focus on the feasible opportunity set and possible activity duration. J Geogr Syst 5(1):71–91
Lucas K (2012) Transport and social exclusion: where are we now? Transp Policy 20:105–113
McNally MG (2000) Four step model. Institute of Transportation Studies, University of California, Irvine, CA
Meyer D, Miller EJ (2001) Urban transport planning: a decision-oriented approach. McGraw-Hill, New York
Morrison K (2005) Structuration theory, habitus and complexity theory: elective affinities or old wine in new bottles? Br J Sociol Educ 26(3):311–326
Pan HX (2006) Shanghai from dense mono-center to organic poly-center urban expansion, Urban Environmental Management Project of IGES, Department of Urban Planning, Tongji University, China
Pan H, Shen Q, Liu C (2011) Transit-oriented development at the urban periphery: insights from a case study in Shanghai, China. Transp Res Rec 2245:95–102
Pei K, Fan D (2010) Passenger flow forecast for urban rail transit based on extended four-step model in Wuxi, China. J Express Rail Traffic 23(5):57–61. (in Chinese)
Preston J, Rajé F (2007) Accessibility, mobility and transport-related social exclusion. J Transp Geogr 15:151–160
Rauws W, De Roo G, Zhang S (2016) Self-organisation and spatial planning: an editorial introduction. Town Plan Rev 87(3):241–251
Rawls J (1971) A theory of justice. Oxford University Press, Oxford
Sen A (1980) Equality of what? Choice, welfare and measurement (Cambridge, MA: Harvard University Press), 1997 [Originally published in The Tanner Lectures on Human Values, Vol. I, by the University of Utah Press and Cambridge University Press, 1980]
Sen A (1984) Resources, values and development. Oxford, Basil Blackwell
Sen A (1992) Inequality reexamined. Oxford University Press, Oxford
Sen A (2000) Social exclusion: concept, application and scrutiny. Social development papers no 1. Office of Environment and Social Development, Asian Development Bank, Manila
Shanghai Statistics Bureau (2017) Shanghai statistical year book 2017, Shanghai Municipal Government (in Chinese)
United Nations (2016) Identifying social inclusion and exclusion. In: Leaving no one behind: the imperative of inclusive development, report on the social situation 2016. Department of Economic and Social Affairs, New York, pp 17–28
Wang D, Gan D, Zhu C, Lu X (2012) A research on the relationships between suburban spatial planning and railway traffic planning in Shanghai. J City Plan 1:17–22. (in Chinese)
Zacharias J (2012) Resisting motorization in Guangzhou. Habitat Int 36:93–100

Glossary

Accessibility The ability of an individual to penetrate constraints to access the mainstream activities of society. The key goal of urban transport planning is increasing accessibility by improving individuals' ability to move around freely and their choices and opportunities in accessing well-being.

Agency The capacity of an individual to act independently and to make his or her own free choices. Also known as action.

Agent A knowledgeable individual, who incorporates, exchanges, creates and exploits knowledge to transform rules and sources to achieve his or her own goals, e.g., employment.

Self-Organisation Process A process that takes place in response to external influences in a social system. A natural and spontaneous mechanism of interactions among individual members in the social system to achieve a state of equilibrium. In this book, external influences are regarded as top-down emergence (structures) and the actions of individuals are regarded as bottom-up emergence (agency) (Fuchs 2003).

Social Exclusion A self-organisation process that produces a state in which individuals are excluded from participating fully in economic, social, political and cultural life in their society. A multi-dimensional phenomenon not limited to mobility and material deprivations but also to the deprivation of freedom and opportunities to access well-being.

Structuration Theory A social theory of the creation and reproduction of social systems (self-organisation) that is based on the cycle of structure and agency. The theory describes the duality of a structure in which the structural properties of a social system are both the medium and the product of agencies they recursively organise. This book argues that the process creates new structural properties that cause accessibility problems and social exclusion.

J. C.-y. Lau, *Self-Organisation Shapes Travel Behaviours and Social Exclusion in Deprived Urban Neighbourhoods of China*,
https://doi.org/10.1007/978-981-15-2252-9

Structures Rules and resources (e.g., government policies, the built environment, norms and social networks). According to Giddens (1984), structures exist only in memory traces, which are drawn upon by individuals in the production and reproduction of routine social actions.

Transit-Oriented Development Compact, mixed-use, pedestrian-friendly development organised around a transit station. Such development embraces the idea that locating amenities, employment, retail shops, and housing around transit hubs promotes transit usage (Cervero et al. 2013).

References

Cervero R, Iuchi K, Suzuki H (2013) Transforming cities with transit: transit and land-use integration for sustainable urban development, urban development series. World Bank

Fuchs C (2003) Structuration theory and self-organization. Syst Pract Action Res 16(2):133–167

Giddens A (1984) The constitution of society. Polity Press, Cambridge

MIX
Papier aus verantwortungsvollen Quellen
Paper from responsible sources
FSC® C105338

If you have any concerns about our products,
you can contact us on
ProductSafety@springernature.com

In case Publisher is established outside the EU,
the EU authorized representative is:
Springer Nature Customer Service Center GmbH
Europaplatz 3, 69115 Heidelberg, Germany

Printed by Libri Plureos GmbH
in Hamburg, Germany